The Boiler Room Boys

The Boiler Room Boys

An Underground Story of Science, Religion,
and the Faith that Fuels Both

Tim D. Smith

Foreword by Francis Capitanio

WIPF & STOCK · Eugene, Oregon

THE BOILER ROOM BOYS
An Underground Story of Science, Religion, and the Faith that Fuels Both

Copyright © 2019 Tim D. Smith. All rights reserved. Except for brief quotations in critical publications or reviews, no part of this book may be reproduced in any manner without prior written permission from the publisher. Write: Permissions, Wipf and Stock Publishers, 199 W. 8th Ave., Suite 3, Eugene, OR 97401.

Wipf & Stock
An Imprint of Wipf and Stock Publishers
199 W. 8th Ave., Suite 3
Eugene, OR 97401

Except where otherwise noted, biblical quotations are from the Holy Bible, English Standard Version® (ESV®), copyright © 2001 by Crossway, a publishing ministry of Good News Publishers. All rights reserved.

www.wipfandstock.com

PAPERBACK ISBN: 978-1-5326-6210-2
HARDCOVER ISBN: 978-1-5326-6211-9
EBOOK ISBN: 978-1-5326-6212-6

Manufactured in the U.S.A.   01/09/19

This book is dedicated to my best friend, my wife and my partner of
more than fifty years,
Margene K. (Sorenson) Smith,
for her long-suffering patience with my time in the boiler room.

Contents

Foreword by Francis Capitanio | ix

Preface | xiii

Acknowledgments | xv

1. The Bomb and the Boiler Room: 1955–1964 | 1
2. Touchstones | 13
3. Growth and Form: 1965–1967 | 19
4. Breaking Down Belief | 29
5. Breaking Down: 1968–1972 | 40
6. Facing the Void | 53
7. Facing Reality: 1973–1997 | 61
8. Our Cosmic Home | 73
9. Coming Home: 1998–2005 | 88
10. Science, Religion, and the Life In-Between | 101
11. Living the In-Between: 2006–2010 | 121
12. Being Human | 131
13. Being Healed: 2011–2018 | 144
14. Freedom to Change | 150
    Epilogue: Back to the Boiler Room | 154

Glossary | 157

Bibliography | 161

Index | 167

Scripture Index | 177

Foreword

I HAVE HAD THE privilege of knowing Dr. Tim Smith for over twenty years. Before my ordination and our move to separate American coasts; before worshipping together in charismatic churches; before working together at NOAA Fisheries; he was a spiritual father to me, caring for me as I came out of a dark and convoluted world. I have always known him as a man comfortable with Boiler Room living. The boiler room, as this book explains, is the place where we ask questions, where we gather information and push forward in a quest for truth, where we may not have the comfort of black and white answers or Christian clichés, but we have the comfort of knowing that, whatever we don't know *now*, the Truth is out there. Dr. Smith has always been, to me, a man of faith and a man of science, and somehow in him the two came together quite naturally. His boiler room began as a real place in the belly of an elementary school, and ours contains the boiling cauldron of angst that modern culture has produced, with its uncertainties and questions and an absence of any tangible Truth.

As the modern Western world rolls on, so does an increase in the complexities and confusion of our collective lives. Technology expands our connectivity, opening windows into private lives that, come to think of it, should perhaps never have been opened. Electronics become the medium through which we hear the clamoring of politics and the so-called culture wars, and the latest tweet and soundbite become fodder for an ongoing fire of activism, discontent, and perplexity. A two-party system unifies us in this one thing: the knowledge that politics, among other hot topics of the day, are destined to divide us. Independents hardly ever make their mark; extremists have the loudest voice; and the average American Joe or Jane gets lost somewhere in the middle. And the world rolls on.

This "choose your side" mentality is not the product of American politics or culture wars. Humans have been dividing themselves into factions with little hope of resolve since time immemorial, since Adam pointed his guilty finger at Eve and said, "The woman made me do it." Moderate, considerate voices have hardly prevailed ever since. In my own spiritual journey, I've often swung between two camps: rationalism and emotionalism. During one season, I questioned almost every notion that I could not see, smell, hear, touch, or taste. Another had me contemplating the secrets of the universe, opening my heart to an invisible God. Books shouted at me from both sides of the political and cultural spectrum. Modern day mystics told me to check my mind at the door and just receive whatever God had for me, while the rationalists spoke without emotion but with conviction: "Eat drink and be merry, for tomorrow we die."

I was a history major in college, and history, like science, is no different in its clamoring sides: an uncivil war of ideas that this happened, this way, and surely, we know exactly how it happened a couple thousand years after the fact. After college, when I entered the workforce among scientists, surrounded by oceanographers and marine biologists, I found myself challenged by many who questioned my growing faith in God and held evolution and its conclusions to be their theology of choice. The discussion was always presented as a rivalry between two irreconcilable factions, something I was used to. That is, until a more moderate voice prevailed.

Clive Staples Lewis was an Anglican writer who, though never an ordained clergyman, served not only the Church of England but all Christians with his many books on philosophy and theology. It was C.S. Lewis' work, *Mere Christianity*, that first got me thinking that perhaps the answer to questions that people in both scientific and religious camps were asking could be found on common ground. He may have become most famous for his Narnia Chronicles, but books like *The Great Divorce*, *Mere Christianity*, and *A Grief Observed* endeavored to bridge the gap between rival philosophical and religious factions. His writing attempted to build a central living space in which people could enter from their various rooms to sit and stay awhile to listen and reason together. They might, in the end, return to their respective quarters, but they will have returned more informed and perhaps more understanding of those who chose different places to sleep. Some might even choose to stay in that living room space, finding it more comfortable in the warm company of others than in the cold isolation of their own great opinions. The fact that a strong Christian thinker like Lewis, who undoubtedly believed in Jesus of Nazareth as Christ and God, could also welcome tension and invite conversation was remarkable indeed. Thankfully, over the decades of the 20th century his thoughtful voice has, on occasion, pulled from both sides individuals who stopped their shouting long enough to hear that still, small voice of moderation and walk into the living room for a lively discussion and a warm cup of tea.

Dr. Tim Smith is another such voice in the tradition of those philosophers, scientists, apologists, and theologians who refuse to maintain an either/or approach to their beliefs and seek to find common ground on which to walk the journey of faith. As I grew to know him as a man of science and of faith, I was inspired to keep having difficult conversations with those around me who, perhaps, couldn't see God through the scientific data. If such a scientist could also be such a devoted Christian, then I knew that I, too, could walk the middle road safely.

The Boiler Room Boys is an attempt to coax some of us out of our neatly classified ways of thinking and into a boiler room where we can ask questions without fear. The boiler room was, for Dr. Smith, a place where exploration could occur freely, where inquiring minds could have a meaningful conversation with each other and with the scientific data. I am thankful to him for his willingness to speak out in the cultural and partisan cacophony around us, to add a voice of moderation to so many voices of division, politicizing, and conflict. It is my hope that those who pick up this book will have their own journey, in whatever camp they're in, informed by his journey and end up finding the boiler room a not-so-scary place to be after all.

The Rev. Francis Capitanio
All Saints Anglican Cathedral
Amesbury, MA 2018

Preface

Conflicts between fundamentalist Christians and fundamentalist scientists left me feeling that I had to choose. My youthful search for truth led me towards science and away from religion. Choosing science gave me security, a career, and a sense of value. Further, I had hoped this choice would help me find answers to the big questions of life. My older brother Eric had encouraged me to ask these big questions, perhaps from too young an age, but neither he nor my science teachers nor my church leaders were able to help me answer them. Although science did not immediately answer those questions, scientists claimed that it eventually would. Decades later it hadn't, and I was left wondering.

Here I describe my long journey into and out of atheism, beginning with a group of boys studying science in the boiler room of our grade school in the mountains of Oregon. My journey included studying religion with the leaders in a community church in my valley. The conflicts between the church and the scientists eroded my confidence in the former. However, my confidence in science was also eroded as the events of my life played out. Eventually, I was left working out my answers to the big questions in my own boiler room.

Here I tell the story of coming to my own answers to science and religion, writing in two streams. In the first stream, the odd-numbered chapters, I have described how conflicts between fundamentalist Christians and fundamentalist scientists led me into and out of Christianity, into atheism, and back to theism and finally back to Christianity. In the second stream, the even-numbered chapters, I have described what I now understand to be the truth about science and about Christianity. In the interaction between these two streams, I expose what has been called a "war between science and religion" to be a false conflict promoted by both some scientists and some Christians.

Preface

My purpose is to convince you that there is so much more to understand about the universe and God than any form of literalism, religious or scientific, can explain. I also wish to convince you that seeking understanding is a doable task, albeit one that will require study in your own boiler room. I am convinced that God loves his people, and will provide a much better foundation for our lives. Finally, I wish to encourage you to be prepared to help those who are asking life's big questions, younger and older, that they would be free to ask and would find truthful counsel that would help them not wander in the wilderness as long as I did.

Acknowledgments

DANN FARRELLY—FOR YOUR HELPING me to understand the change God made in my life, and how that relates to both my science and my faith.

Franklin Capitanio—for your helping me to think out and write out my faith.

Vanessa Chandler—for your early guidance on how to organize this book.

My students over the years—for your patient responses to my story and my teaching in both science and religion.

1

The Bomb and the Boiler Room: 1955–1964

FISHING POLE IN HAND that Sunday morning, I blushed crimson when Mrs. Lane wagged her finger at us from her car window. I was walking with my brother Eric past a tiny church in the Cascade Mountains of Oregon situated on the banks of Camp Creek, near where it feeds into the McKenzie River. There Missy, as she liked to be called, was my sometimes Sunday school teacher.

"Why am I blushing?" I demanded of myself, embarrassed at my embarrassment. After all, we didn't have to go to Sunday school. Pop and Mom had given us a gift of skepticism about God and churches. They were rebuilding lives twice interrupted as they finished high school, first when Pop was shipped to Europe to fight the Nazis and then when he was shipped to the Philippines to fight the Japanese. Their lives had been interrupted but not shattered, although they never reconciled their war experiences and their ideas of God.

The start of the fishing season seemed a good excuse to skip Sunday school. I was increasingly finding Missy's teaching unconvincing. Also, my brother Eric, four years older than my 12 years, had been challenging me about what he called "big questions." Lately, we had been arguing about the origin of the universe and of life. His high school was on the other side of the McKenzie River, and his science teacher there said that the universe and the earth had always been here just as they are. Further, someone had recently created life in the laboratory. Missy on this side of the river claimed that God created the universe and everything in it 6,000 years ago, including life, and everything would soon be gone when Jesus returned. I made Eric angry when I quoted Missy and Missy angry when I quoted my brother. Easier to go fishing in the river flowing between these two views.

That year I was in the sixth grade at Camp Creek Elementary School. There had only been six of us since the first grade—all boys. We shared an overcrowded classroom with the much more numerous fifth graders, and the teacher had a hard time keeping up. He was more than happy to let some of us study science by ourselves in the room with the school's diesel-fired hot water heater, the windowless boiler room. We mostly ignored our dreary science textbook and started looking for other things to study, competing to find the most interesting topics. We called ourselves the Boiler Room Boys.

I became the leader because I was oldest by nearly a year. The age difference was due to a birthday anomaly that had delayed my starting first grade until I was almost seven years old. The then-new school superintendent had suddenly begun enforcing a long-ignored birth date cutoff rule. My Christmas birthday meant I wasn't going to start first grade the fall I had expected, but instead would have to wait a full year. I was incensed. Although I challenged my parents to protest this injustice, they either didn't or failed, I never knew. During that year of waiting, I often relived my anger as I played with my toy trucks and my favorite airplane, a B-29 Superfortress, alone on the dirt hillside in front of our home.

Pop had said that this type of airplane had dropped atomic bombs on Japan to end WWII, bringing him home sooner from the Philippines. Unbeknownst to me then, B-29s had just begun bombing raids over Korea.

We Boiler Room Boys copied a periodic chart of the elements from a chemistry book onto a large piece of cardboard. This iconic image of chemistry had been created in 1849 by a less than promising student, Dmitri Ivanovich Mendeleev (1834–1907). He was raised in his grandfather's Russian Orthodox Church, but after he remarried, he adopted deistic beliefs in part because his church forbade second marriages. Mendeleev became a chemist and organized the 63 then-known chemical elements by their atomic weight and chemical properties. He reportedly created this using his fascination with a card game named Patience, one that we knew as Solitaire. Mendeleev studied the patterns among the chemical properties of the elements to conjecture the existence of seven then-unknown elements. Eventually, three of those predicted elements were discovered, validating the power of the patterns in his chart, but also demonstrating that such patterns weren't always reliable.

The Bomb and the Boiler Room: 1955–1964

We especially wondered at the radioactive elements on the chart, each of which we marked on our chart with a carefully drawn radioactivity symbol. One was uranium, an element that has since colored every aspect of my life. We learned that scientists at Los Alamos, New Mexico, had used uranium-235 and plutonium-239 to make the war-ending bombs that American B-29 Superfortress bombers dropped on Hiroshima and Nagasaki. By 1949 Russian scientists had also exploded a plutonium bomb, the reason for our frequent air raid drills. During those drills, we were to hide under our orderly wooden desks. We watched a film called *Duck and Cover*[1] about what to do if, but seemingly when, Russia dropped atomic bombs on us. The film attempted to draw lessons from pictures and films of the devastation in Japan caused by our atomic bombing and to reinforce those lessons using a lilting song about a cartoon turtle ducking and covering in its protective shell. The devastation of Hiroshima was seared into my mind, despite the reassurance that those who survived the blast and the radiation burns soon returned to their normal lives. The Boiler Room Boys thought that this duck and cover drill would only help make our bodies easier to identify. But we quickly learned that our teacher didn't want to hear that; there was no money to make the increasingly popular bomb shelters for school children in the remote valleys of Oregon.

We found the astronomer that Eric's science teacher had quoted, sometimes science fiction writer Fred Hoyle (1915–2001). He claimed the universe was infinitely old and in a "steady state" with its continuous creation of matter. And we learned about an experiment to create life conducted by Stanley Miller (1930–2007). He shot electricity into a hot methane and ammonia soup, thought then to mimic conditions on the early earth. His experiment created amino acids[2] but was often reported as having created life. These were the things of science, and we assumed that they were true, just as Mendeleev's periodic table of the elements was true. Science seemed to be able to answer some of Eric's questions.

The Boiler Room Boys also reached out into philosophy. I began putting up glossy folded charts from the nearby university's bookstore that explained philosophers' philosophies. The names and pictures of the philosophers were printed down the wide pages, and the characteristics of their philosophies were stretched across the pages. Perhaps here the tension

---

1. Mauer, *Duck and Cover* (1952).
2. Miller, "Production of Amino Acids."

between Missy's world and what I increasingly saw as my world, science, could be resolved.

Of course, I didn't really understand the charts, but I liked the names, especially the strong-sounding ones: Kant, Spinoza, Hume, Comte, Plato, Socrates, Euclid, Aristotle, Marx and many more. And I was encouraged just by reading the names of their philosophies: positivism, nihilism, dialectical materialism, existentialism and, again, so many more. I was sure that when I understood the truth of all of these philosophies taken together, I would find answers to the big questions Eric kept encouraging me to ask.

Most of the conclusions of the Boiler Room Boys were inconsistent with Missy's Sunday school teaching. Jesus wasn't listed among the philosophers; the earth had always been; life was easily created. In Missy's world, God created the earth, life, and us; we sinned; and Jesus died for our sins so we would love each other.

Yet Jesus didn't seem important because people didn't seem to love each other. For example, we were aware that many civilians had been killed in World War II, Jews in Germany by the Nazis, Poles in Poland by the Soviets, Germans in Germany by our firebombs using the newly invented napalm, and Japanese in Japan by both our firebombs and nuclear bombs. I remember hearing that killing millions of Jews was justified because of who they were, murderers of Jesus. Further, our killing hundreds of thousands of Japanese civilians in an instant was justified because it had been calculated that stopping the war with atomic bombs would prevent the loss of many American soldiers' lives during a land invasion, possibly my father's life and hence my existence.

Such justifying calculations sounded so scientific, but I wasn't quite convinced they were true. They must depend on understanding the value of people, and I already knew that science did not deal well with questions of value. I did not yet know that people valued their own people more than those of other groups, despite what Missy said Jesus taught, to love your neighbors. And I did not yet know how to do such calculations.

One cold day in the spring our teacher visited us Boiler Room Boys. He seemed interested in our philosophy charts but then frowned, pointing at one of the philosophers.

"Didn't you know," he demanded angrily, "that Karl Marx's ideas were wrong and led to the Russian Revolution, and that our air raid drills were because Russia also had an A-bomb?"

The Bomb and the Boiler Room: 1955–1964

We hadn't made that connection between Marx and air raid drills, and I was surprised at the teacher's anger.

"Marx's philosophy was wrong," he nearly shouted, "and you shouldn't have his picture up there."

This bothered me. I had assumed that all the philosophies were true, just emphasizing different things. If Marx had been wrong, what of the others? My hopes began to fade that philosophy would reconcile Missy's world with that of the Boiler Room Boys.

The next week our teacher returned to our Boiler Room but this time with a man in a uniform. We were clustered around our work table, the philosophy posters between us and the hot boiler. Our teacher looked nervously at the uniformed man. The badge on his dark blue uniform read "Fire Marshal," and his scowl and the golden braid looped around his shoulder said he meant business.

"There is a problem," the fire marshal said grimly, looking aggressively at the teacher and then at us. "How long have you students been meeting in this room," he continued.

"Since last Se—" I started to answer when our teacher interrupted.

"Just a few —" he stammered before the fire marshal cut him off.

"I need to hear the students' answer," he continued. "I've heard yours."

I didn't know what to think, but it was dawning on me that the original question wasn't a small matter for the marshal, and he seemed to be accusing our teacher of something bad. I realized that while our teacher had been worried about a picture of Karl Marx, the marshal was only worried about fire hazards.

"We've only had the posters up for a few weeks," I lied to divert the marshal from his original question. "Are they a fire hazard?"

"A fire hazard?" the fire marshal shouted. "Isn't it obvious. Whatever you have been doing in here, you must stop it at once."

"We're sorry," our teacher said, wincingly. "They won't be allowed in here any longer." We collected our periodic chart of the elements, by then tucked up against the boiler and hot to the touch, and our philosophy posters and books. Our time of asking whatever questions we wanted from the safety of the Boiler Room was over.

While we weren't forbidden from studying Karl Marx, our teacher effectively stopped us by bringing the fire marshal, all in the name of our safety. This also solved our teacher's concern about us having a picture of Marx on the wall. I was afraid to bring up the philosophers when we were back in the classroom. The Boiler Room Boys gradually disbanded as we moved on to junior high school on the other side of the river.

Calculations of War

In junior high, I became enamored with the ideas of mathematics and the possibility of learning what was true using those ideas. I was now riding the school bus 20 miles out of our little valley early each morning, and the world began opening. Soon I was taking classes in algebra, which I learned was created by the Persian Khwarizmi, and geometry, created by the Greek Euclid. I remembered the name Euclid from my philosophical posters, but now Euclid the mathematician stepped in front of Euclid the philosopher. My mathematics teacher described how the word *algebra* had come from the full name of the Persian astronomer and mathematician Al-Khwarizmi (died AD 850). His book was titled *The Science of Restoring What Is Missing and Equating Like with Like*, and I hoped that his secrets of manipulating Xs and Ys would reassure me that the calculations I hoped were behind the decision to bomb Hiroshima were, in fact, possible.

These mathematics classes also led me further from the murky world of the philosophy charts now pinned up on my bedroom wall. I began seeing another way to find the truth: logic. Deductive logic was fascinating and using it I could deduce the seemingly never-ending theorems of Euclidean geometry from a few assumptions or axioms. The axioms were simple and seemed obvious, and the theorems, when accurately deduced, felt true. My teacher loved calling students to the blackboard to prove theorems. One day she called on me:

"So, Tim, would you demonstrate that the interior angles of a triangle always add up to 180 degrees?"

I went to the board and groped my way to complete the proof. I was both terrified that I couldn't and exhilarated when I did. This wasn't the groping for patterns that Mendeleev had used in deriving the periodic chart of the elements. I realized that I could have used that approach, for example, by patiently measuring the angles of many triangles. But how many triangles would I have to examine to assure myself that the angles of all triangles added to exactly 180 degrees? Now deduction proved to me that this theorem was true, not just a matter of conjecture or belief or approximate truth. Euclid's deductive logic was preferable to Mendeleev's patterns, or inductive logic, because it produced true theorems rather than just predictions of possibilities.

I soon learned that Euclid's deductive logic was like other scientific developments in being based on mathematics. Physicist Isaac Newton (1643–1727), an unorthodox Christian who thought that one could only come to God through the "frame of nature,"[3] had derived the elliptical motion of the planets based on his theory of gravitation and on the three laws of motion developed earlier by the pious Lutheran[4] Johannes Kepler (1571–1630).

Later, deistic[5] physicist Albert Einstein (1879–1955) challenged Newton's model. There was no absolute motion, and, in fact, motion could only be detected relative to something else. I remember one afternoon trying to prove to our mathematics club using Einstein's equations why we could never travel faster than the speed of light. I didn't understand his theory, but everyone could see the problem when zero showed up in the denominator. You can't divide by zero, but this was actually not a proof of anything. I merely demonstrated that Einstein's equations were consistent with what he had assumed, that the speed of light was a constant.

I was a little concerned, however, that Newton's truth was contradicted by Einstein's. If so, I wondered, could the calculations that underlay the B-29 Superfortress bombing of Hiroshima also be wrong? The differences between Newton's and Einstein's calculations were small in the practical world, so I ignored them, but the differences in the relative values of people continued to haunt me.

Then on May 1, 1960, my questions about truth and reality abruptly changed. That was the day the United States government announced that the Russians had shot down an American weather plane that had "innocently" flown off course over Russian territory. I was outraged until Russian Premier Khrushchev displayed the wreckage of Lockheed's U 2 airplane, a very high-altitude photographic reconnaissance airplane named the "Dragon Lady." President Eisenhower had been lying, and Premier Khrushchev had been telling the truth. If Eisenhower was lying, I began wondering about other political issues. I wondered again about the calculation justifying the bombing of Hiroshima. I had wondered about the difficulty of

---

3. Hummel, *Galileo Connection*; Iliffe, *Priest of Nature*.

4. Kepler said in 1595, "Behold how through my effort God is being celebrated in astronomy." Quoted in Wertheim, "God Is also a Cosmologist."

5 . In a talk at the Union Theological Seminary in the 1940s Einstein said, "Science without religion is lame, religion without science is blind." Quoted in Isaacson, *Einstein*.

even doing such calculations, but now I wondered about the truthfulness of those claiming to have made them.

Soon after the U-2 spy plane crashed, I asked my history teacher if we might have a class on Russian history next year. I was surprised to find myself immediately marched to the guidance counselor's office despite my protests that I could find my way. I sat in the hallway while my teacher met with the counselor. Then it was my turn.

The counselor began asking questions about why I was interested in Russia, and I explained about the U-2 spy plane and Eisenhower lying. I described our grade school teacher shouting about Marx and the Russians having a plutonium bomb and the *Duck and Cover* film. She seemed not to know about Eisenhower's lying or about Marx's philosophy. She continued asking questions for a half an hour or so.

I thought I made a good case for studying Russian history. Then she sat back and, letting out a deep breath, asked, "Tim, wouldn't a class in Canadian history be more interesting? It is much closer than Russia, after all." I realized that she was the fire marshal all over again, trying to protect me, or someone, from my request. I did not get a Russian history class, but she did put a note about my request on my "permanent record," a note that would haunt me for several years.

Science Fair Fraud

I rushed out to the greenhouse when I got off the school bus that Friday afternoon in April. Dread filled me: the science teacher had reminded us that the science fair would be on Monday. He seemed to shout this fact at me: "It is mandatory for all sophomores, so bring your experiment or get an F."

When the science fair had been announced the previous November, I remember being excited. We were to do an experiment of our choice, and I immediately thought that I could do something in the family's greenhouse. I had been fiddling with the plant growth hormone gibberellic acid and was sure I could do something with that.

As I read the science fair instructions, however, I was taken aback. It began with a long, tedious definition of the scientific method, followed by detailed instructions for making hypotheses and setting up experiments. It described how to collect and record data and compare results. It even described at length how to phrase your answers. I already knew how to set up experiments, I thought, and this seemed to be written for grade-school children. I bristled at being talked down to, and I again felt my first-grade anger at the school superintendent. I had immediately vowed then not to do a science fair project that year.

That April Friday, however, my vow failed. Science would be my career, I was pretty sure, and an F in a science class was not acceptable. On the long bus ride home, I thought about how to manufacture a science fair project. I remembered that I had sprayed growth hormone on some tomato plants earlier that spring and was relieved to find them still alive in the far corner of our somewhat decrepit greenhouse. I was even more relieved to find some other untreated tomato plants in another corner. Although I had just been messing around with plant hormones, perhaps I could manufacture a science fair experiment after all.

I unearthed the science fair rules that had put me off earlier to make sure I had what was required. Control and treatment groups? Yes, the two groups of tomato plants could qualify. Recorded measurements? Yes, I could still make out my crude notes written using a broad carpenter's pencil on the pine greenhouse benches; the dates were unclear, however, so I'd have to interpret them. A hypothesis? Well, not so clear. I had just been interested in what the hormones would do, fascinated by how I could manipulate the tomato plants at my will. Drawing a simple conclusion from the "experiment" wasn't so obvious either. Although the hormone-treated plants had zoomed ahead, now all the plants were about 18 inches high. All the plants' leaves were beginning to yellow.

That weekend I did manufacture a passable science fair experiment out of the leftovers in the greenhouse. A fancy backdrop, a partially true project description, and then an explanation. I credited the lack of overall effect of the hormones to the experimental conditions, namely the plants becoming pot-bound, so the slower untreated plants eventually caught up with the treated plants.

On Monday I took the backdrop, description, and the treatment and control groups of tomato plants to school, but blushed that afternoon when the principal called me to the front of the gymnasium to receive a first-place award. Although I had often blushed in shyness then, now I blushed in shame. My "experiment" was a fraud regardless of how I dressed it up.

My teacher wanted me to take my experiment to the regional science fair the next week. The instructions said that there I would have a face to face interview with the judges, an interview that would require more deceit.

I confessed to my teacher how I had manufactured my project at the last minute, but he seemed more concerned about our school being represented than with my honesty.

"You illustrated the scientific method so clearly," he said. "It is not important how you got the work done because the truth of science is independent of the scientist. You'll still get an A grade."

This was an attractive argument and would have allowed me to represent my school and myself regionally. After all that was said and done, the results of my "experiment" were nonetheless true, that is, I had made these observations. I hadn't conducted it exactly as I had described, I argued with myself, but what difference did that make? However, I still felt like the science teacher was asking me to explicitly lie to the regional judges just as I had implicitly deceived the local judges.

That evening I took the project description and fancy cardboard backdrop out behind the greenhouse to the family burn pile, and then I planted the tomatoes in our garden. I didn't know what else to do because I suspected the teacher was wrong and because I was ashamed. Just as President Eisenhower had lied about our spy plane, I had fabricated a science experiment. My experiment had been a lie, but my teacher had encouraged me to act like it wasn't. It was my choice to destroy the project, and I acted like the fire marshal to protect me from being exposed.

Collecting Animals

By my junior year of high school, I was allowed a self-directed biology class. I cultured slime mold in the laboratory on baking sheets in a heated rack and learned how to cause the mold to shift in a few days from single cells to multicelled forms, called "dog vomit fungus." I was intrigued by how the individual cells coordinated themselves to form a mass of cells and, in that, to produce a fruiting stalk. It was like controlling the growth of my ill-fated tomato plants; I liked it.

I went with my teacher and a group from a nearby university collecting lizards in the high desert in eastern Oregon. There we collected western fence lizards by stunning them with large rubber bands hooked over our thumbs and shot with our fingers. We quickly gained accuracy, collecting ever-increasing numbers of lizards. The study I eventually learned was to describe the fence lizard population, especially color patterns.

The lizards would mostly wake up from being stunned, and we soon had a box of many dazed lizards. Those that didn't wake up were dumped in the camp garbage cans. A couple of days later, when we broke camp, all the remaining still-living lizards were also just dumped into the cans. I was appalled, nearly crying, but did not speak out. I could not ask about their actions because these were real scientists. I didn't want to embarrass myself.

I also studied skeletons, boiling the meat off to display the skeletons within. The skulls especially fascinated me. I had a good skunk skull I got when we tore a barn down and several deer skulls from hunting trips but didn't have a rabbit skull. One day I took Pop's 12-gauge shotgun and my hunting knife and went up on the hill behind our farm. I was used to hunting deer; this would be simple.

The rabbit startled me when it broke cover, dodging through the thick grass first left and then right. I pulled down on it, expecting it to go left again, and squeezed the trigger, hoping for a clean kill. But the rabbit went right instead, and unexpected tears spurted from my eyes when I saw him lying there badly wounded, breathing hard with fear in its brown eyes. I cut its throat, but I couldn't stop my tears. It didn't die easily, and when its heart had nothing left to pump, something disappeared, something I didn't understand and couldn't quite see but that I knew was gone.

Answering the "Big Questions"

Science's answers to many of the big questions that my brother Eric and I had discussed seemed sensible to me, but I was embarrassed that my church often had different answers.

For example, scientists claimed the universe had no beginning and had always been just as we see it, while my church claimed that the universe had had a relatively recent beginning. Similarly, scientists claimed that they could create life along the lines of Miller's 1953 experiment. In contrast, my church claimed that all life was created just as it is today.

But there were many more questions where science's answers were less clear. What disappeared when my rabbit died? Science claimed that nothing really happened, while my church claimed that at least when people died their "spirit" continued to exist. But for rabbits, the church's answers were less clear.

"What is true?" was the question that perplexed me the most. My brother had often posed this question about the nature of reality. "What is really out there, outside ourselves?" was a favorite question of his. In Sunday school we had memorized John 8:32, where Jesus seemed to be saying that when you study the Bible "you will know the truth, and the truth will set you free." That is, everything you need to know is in the Bible. In science classes the claim was more modest: we don't know everything, but science is a method that will eventually reveal all truth.

A different question that my brother often asked was if it was always important to tell the truth. I remember memorizing Ephesians 4:25: "Therefore, having put away falsehood, let each one of you speak the truth with his neighbor . . . " However, my science teacher encouraging me to continue to lie about my science fair project suggested that science was less demanding on truth-telling.

Similarly, I had seen that church members sometimes seemed to be telling different stories to different people. I knew that I had lied to the fire marshal when he discovered the philosophy posters the Boiler Room Boys had hung around the boiler. But that hadn't seemed so bad, especially compared to President Eisenhower's lie to Premier Khrushchev about the US spy plane. I knew that after I had first lied about my science fair project I did not feel right, wondering if I would be found out, and I wondered about what Eisenhower had felt when he was found out by Khrushchev.

Both religion and science claimed to have answers to these big questions, but in fact they weren't the same answers, and I couldn't reconcile their inconsistency. That left me in a dilemma: I couldn't tell whether science or religion would best help me answer the big questions, but I felt I had to choose one or the other. By the time I chose a college I had turned increasingly towards science. Plus, science didn't lie, and even when it seemed to falter, science offered the hope that it would eventually be able to answer such questions. What I wasn't prepared for were the untested touchstone beliefs embedded so deeply within it.

2

Touchstones

Looking back, I see that there were underlying problems with how religion and science arrived at their answers. Those problems included untested touchstone beliefs that are themselves used to test conclusions about the world. The Merriam-Webster dictionary describes a touchstone as a piece of slate formerly used to test the purity of gold, and thereby metaphorically defines a touchstone as "a test or criterion for determining the quality or genuineness of a thing." Science teachers of my era tended to have an overly simplistic definition of science that didn't recognize those underlying criteria upon which scientific exploration is actually built.

The definition of science is contentious, something that I should have seen when my science fair project went askew. One dictionary understanding of science is "a system of knowledge covering general truths or the operation of general laws especially as obtained and tested through the scientific method."[1] That definition relies on a definition of "the scientific method," the bugaboo question asked of high school science fair contestants. And that definition leads the unwary into defining concepts such as experiments and testing hypotheses. The complete dictionary definition overall runs to several paragraphs.

Despite the complexity of such definitions, many science practitioners like the idea that they are contributors to a system of knowledge.[2] This is often incorporated into many exclusive "societies of learned persons," elected by their own members, such as the National Academy of Sciences,

---

1. There are many definitions of science; I've used Merriam-Webster's online dictionary here and more generally: http://www.merriam-webster.com/dictionary/science.

2. Dalrymple, *Ancient Earth*, ch. 1.

authorized by President Abraham Lincoln in 1863 and often seen as "a body of established opinion widely accepted as authoritative."

This approach to defining science is flawed, however, because it leads to the concept of "scientific" versus, pejoratively, "nonscientific" thinking. Thus, scientific thinking has usually been reserved for those who have bona fide credentials (candidly, an earned PhD degree). Such credentials give practitioners authority in interpreting observations that have been made on almost any topic, regardless of the discipline in which they were trained.

There are, however, other approaches to defining science, some that avoid forming "a system of knowledge." For example, C. John Collins (1954–) suggested defining science as many "discipline[s] in which one studies features of the world around us and tries to describe his observations systematically and critically."[3] Although the disciplinary practitioner pursues his observations of nature based on his own understanding, this definition doesn't necessitate having an exclusive credentialed group of scientists to make or approve of such interpretations. This approach leads to the much more satisfying distinction of logical versus nonlogical thinking in interpreting the disciplinary observations.

Interpretation in this approach is not limited to a self- or culturally appointed elite but becomes a free-for-all, including practitioners of all disciplines, of course, but also philosophers of all persuasions, theologians of all beliefs, logicians and mathematicians of all stripes, and artists of all visions. Anyone who will expend the energy to understand the observations being made and employ logical thinking to evaluate plausible interpretations, not easy tasks, would be welcome. The criterion for truth becomes logic, not pronouncements by priests, neither those ordained by churches nor those ordained by academia.

Looking back at my career in science, this approach to defining science fits the reality of how science is actually practiced much better than the science fair definition.

Types of Observations

Repeatable observations are especially valuable when sufficient details of what was actually observed are reported. One well-known type of observation is the results of experiments constructed to compare the effects of varying one or a few conditions thought to affect the outcome. We depend

---

3. Collins, *Science and Faith*.

on such experiments to determine the efficacy of chemical treatments, for example, when comparing the growth rates of tomato plants that have been treated with growth hormones or the survival rates of cancer patients treated with experimental drugs.

Patterns in observations are also useful, as exemplified by Mendeleev's periodic table, which had fascinated me, along with the other Boiler Room Boys, in grade school. The patterns that Mendeleev saw in his array of chemical elements suggested to him, through the logical process called induction, the existence of some additional although then-unknown elements. Such patterns suggest but do not themselves prove anything, as seen in the fact that some of Mendeleev's predicted elements have been confirmed while others have been shown not to exist.

The seemingly most reliable conclusions are deduced from certain starting assumptions.

A classic example is Euclidean geometry, where for example triangles can be shown to contain exactly as many degrees as half a circle. While such properties of shapes could be suggested using induction based on patterns obtained from examining many such triangles, much stronger conclusions could be possible using deduction. Such conclusions depend entirely on the assumptions made; in this case, that the triangles are flat rather than, for example, spherical. The validity of such assumptions is always a matter of conjecture and knowing if they fit reality is often difficult.

Interpreting Observations

The above types of observations, repeatable inductions from patterns and deductions, have been used in many ways to learn more. I first remember feeling joy with mathematics when I figured out my first proof in Euclidean geometry in junior high school. Something fell into place inside me, and I felt satisfied, excited about "getting" deductions. That satisfaction led me from Euclid's geometry to algebra, trigonometry, calculus, differential equations, Boolean algebra, and statistics, each one proceeding from or branching from the others. I have been fascinated ever sense of how logic and mathematics are so useful in describing the physical and biological worlds.

I remember feeling not quite so satisfied with Mendeleev's inductive analyses that identified gaps in his table of elements. Logical deductions can be derived from premises, but inductions could only be seen in the patterns. The validity of deductive and inductive logic in geometry and chemistry can be tested by repetitive experiments. For example, the implications of Euclid's logic have been tested by repeated observations that the interior angles of a triangle do sum to 180 degrees. Similarly, Mendeleev's prediction of yet unknown elements was tested by experiments when most but not all the predicted elements were ultimately found.

There is, however, another type of observation, those that logically cannot be repeated. One example was the observation by Chinese cosmologists of a "guest star" that appeared suddenly in AD 185 near the star Alpha Centauri. They reported that the brilliant star was visible for eight months and then disappeared. That observation was not the result of an experiment, and, of course, could never be repeated. But it was nonetheless valuable because it was reported in sufficient detail to allow confirmation using modern telescopes, which have detected remnants of the explosion of a star near Alpha Centauri some 8,000 light years from the earth.

Intrigued by how we study nonrepeatable events, the logician Charles Sanders Peirce (1839–1914) in the late nineteenth century defined the logic of "guessing," which has been dressed up with a dignified title, "abductive logic." Peirce thought that arguments about nonrepeatable events fundamentally involved guessing possibilities, which, if true, would be consistent with some "surprising fact" that already had been observed or could be observed in the future. In the case of the Chinese observation of the "guest star," one of the many guesses was that some explosion had occurred. The "surprising fact" that gave cosmologists "reason to suspect" that this was true was the recent detection of cosmic debris near Alpha Centauri using modern telescopes. Not a deduction and not an induction from a pattern, but a guess that was eventually found to be consistent with other observations.

Applying Collins's Definition of Science

The logical, mathematical, and philosophical tools that Collins identified provide a firm basis for interpreting our observations of the universe and the world, both doing science and doing theology. In addition to those tools, however, Collins emphasizes that there are many "touchstone" beliefs that we adopt in order to use these tools, for example, that we exist.[4] We

---

4. Ibid., 22.

will take up the importance of such truths in chapter 4 under the concept of worldviews.

But a more concrete example of a touchstone belief that has influenced cosmology is that the universe is unchanging. Holding this belief affected the development of our understanding of the basic nature of the universe, and it was held for centuries despite the observation of the "guest star." Surely we see now that the appearance of a bright star for eight months implied the universe was changing, but cosmologists up into the twentieth century couldn't see this because it was contrary to their belief that the universe was not changing.

It wasn't until 1925 that the unchanging nature of the universe became a problematical belief when the cosmologist Edwin Hubble (1889–1953) showed that many of the galaxies he observed were moving away from the earth at an ever-increasing speed. That is, the universe was changing. His moving galaxies were a "surprising fact" that was inconsistent with the long-standing assumption of an unchanging universe. In terms of Peirce's abductive logic, the implication can be spelled out as: "If the universe were expanding, then Hubble's observation that galaxies are moving away would be 'a matter of course.'"

That cosmologists believed in an unchanging universe long after the evidence in retrospect at least was clear. Not only did that delay the development of cosmology but it also caused the biblical claim in Genesis and elsewhere of a changing universe to be rejected out of hand. At the very least, that difference affected the relationship between religion and science negatively. We will consider the development of cosmology further in a later chapter.

Lest this problem with cosmology appear as an isolated incident, geologists long held that the earth had also been unchanging. This belief affected how they interpreted geological observations, for example, the layers in the soil and the fossil sea shells on mountain tops. However, geologist Alfred Wegener (1880–1930) in the early twentieth century argued that the continents of Africa and America looked as if they had been pulled apart. Only after World War II did geologists recognize that the world had been changing, allowing geology to develop. As per cosmology, though, the persistence of the assumption the earth had not been changing delayed the science. It also contributed to conflicts between geology and biblical claims in Genesis and elsewhere, for instance about interpreting the flood story.

The acceptance of an unchanging universe and unchanging continents are examples of giving insufficient attention to the touchstones that scientists hold. This acceptance resulted in apparent conflicts across disciplines, including between science and religion. Collins's definition of science recognizes that touchstones are just assumptions and encourages people from many disciplines to engage in interpreting the observations that are made using a wider variety of touchstones. In effect, the success of different touchstone at explaining our observations serves as a test of the touchstones themselves. As I would soon discover, however, scientists aren't the only people struggling to examine the touchstones of their beliefs.

3

Growth and Form: 1965–1967

ON AUGUST 2, 1964, just as I was preparing for my senior year in high school, four North Vietnamese sailors were killed in a brief sea battle. Although at the time I didn't notice this "Gulf of Tonkin affair," it would prove to have a tremendous impact on me and more so on many others in my generation. The next spring, I began the rituals of finishing high school. I took college placement tests, began applying to universities, and registered for the military draft. I had no reason to suspect the surprises that were coming.

The first surprise was that, although my test scores were excellent, the nearby University of Oregon advised me not to bother to apply.

"You have an excellent application," the college recruiter began, "but there are so many students applying this year that we state schools won't be giving many scholarships."

My jaw dropped.

"But I need a scholarship. We don't have the money for the tuition, and I've worked so hard to earn a scholarship," I said, shaken.

"Well, yes," the recruiter replied, "I understand. But there are many more students finishing high school this year, those born just after the end of World War II. And this is just the beginning of a big baby boom. Too bad you didn't finish high school last year."

His comment reminded me that I would have finished high school the previous year if the school superintendent hadn't begun enforcing the birthday cutoff rule when I was five years old. I wondered if that arbitrary decision would keep me out of college.

As a seeming afterthought, he added, "You might try a smaller private college."

Another surprise occurred when I applied for a student deferment from the military draft I had just registered for. College draft deferments had been routinely given by the Selective Service System, but I instead I got a letter from the local draft board asking for more specific information about my college plans. Even though there were more young men available because of the baby boom, more draftees were expected to be needed because of the Gulf of Tonkin affair. I wondered if four North Vietnamese sailors would keep me out of college even if I could get a scholarship.

These problems were resolved by my girlfriend, Margene Sorenson. I had first seen her when we were in an American history class the year before. She was new to my school, whip-smart, and pretty; we began arguing in class debates almost immediately. She seemed taller than she was.

Eventually, I invited her to go bowling, there being little else besides movies in our town, and was surprised when she said yes. But that night the old ladies' bowling leagues had taken over the only bowling alley, and we ended up at the pool hall down the street.

She handled the pool cue well, getting a good opening break. I noticed her long red fingernails complementing her red hair.

"What do you want to do after high school?" I asked after a couple of games.

"I want to go to Pacific Lutheran University and then be a medical missionary to China," she replied immediately, sounding edgy.

I was surprised that she had such definite plans, impressed. I had never heard of that university.

"You'll not be able to keep those pretty fingernails if you do that," I responded, a little aggressively. What was I doing? I wondered.

"And you," she snapped back. "What do you want to do?"

All I knew was that I wanted to go to college and maybe become a scientist and especially to not live in this little town. I was a little embarrassed that I couldn't be more specific.

When the college recruiters came, she interviewed with the one from Pacific Lutheran University. She said that it was a good school and had a good biology program; maybe I should just talk with the recruiter. Reluctantly I did, and I got a scholarship and a promise of participating in one of their biology department's summer research programs. And with that admission, I also got a deferment from the draft board. My fears fell by the wayside, and Margene and I were off to school together, one of the best decisions I ever made.

We began college that fall, and I focused on biology and mathematics. The biology and mathematics lecture classes and the biology field trips were invigorating. The biology laboratory classes, however, were tedious and boring; everything must be arranged *just so*. But I loved the library. I remember climbing the stone stairs of the library in old Xavier Hall to the top floor where there were cozy study carrels and archives of old books. The treads of the first steps leading into the building were irregular, the soft stone worn with deep grooves. The grooves were less pronounced on the second floor and especially on the top floor. We would rush there after dinner, my now girlfriend Margene and I, to claim two of the coziest carrels wedged under the sloping roof.

I had failed to read the biology department's summer reading assignments before classes began. One was Julian Huxley's (1887–1975) *Evolution: The Modern Synthesis*. I knew a little about Charles Darwin's (1809–1882) theory of evolution but was worried that I didn't know enough and was expecting to be asked about my summer reading any time now. I was concerned as well because I had seen in Julian Huxley's table of contents the phrase "The Eclipse of Darwinism," and so began reading that book in earnest.

I had learned in high school that life had emerged from amino acids created by electricity striking a warm pond and that all organisms had diversified and evolved from that first life by natural selection. The teacher had especially emphasized that Darwinism had triumphed over creationism. Darwin's natural selection was simple enough at least in broad strokes. Individuals vary but usually look mostly like their parents. Some of these varying offspring would tend to reproduce more than others so that gradually more and more individuals in the population would come to be like those who reproduced more. If conditions changed, the population of organisms would gradually begin to change, generation after generation, with individuals with characteristics that were more useful under the new conditions becoming more prevalent.

Julian Huxley's phrase "eclipse of Darwinism" sounded like evolution had been disproven, which Missy Lane at the Camp Creek Community Church would have loved to hear. But it wasn't evolution that had been disproven. Rather, Julian Huxley was just describing how Darwin's original ideas about natural selection were rejected by biologists over the 50 years or so after he published his *Origin of Species* in 1859. Darwin's idea had failed to impress many biologists then, especially because they and Darwin had continued to believe in blending inheritance, where the differences among parents were averaged out in their offspring. I knew immediately that this contradicted Darwin's natural selection, and so did most biologists at the end of the nineteenth century.

Reading on, however, I was relieved to find, despite this initial rejection of Darwin's theory, that Huxley still felt that all life had evolved from simpler life because the monk Gregor Mendel (1822–1884) had developed a better theory of inheritance. Thus, my basic belief that biology did not require God to explain the origin and diversity of life still sounded true.

Darwin's theory had gone into decline because he had failed to embrace Mendel's theory, which Julian Huxley saw as eventually rehabilitating Darwin. He had not embraced Mendel's theory perhaps because he never became aware of it, but more likely because Mendel used mathematics to interpret his very careful experiments with pea plants. Darwin stated in his autobiography that mathematics "was repugnant to me," and was "like a scalpel in a carpenter's shop—there was no use for it."

Although the biology department never showed any interest in its summer reading assignments, there were several topics that I continued to investigate on my own. These included how mathematics related to biology and the definition of life.

D'Arcy Thompson's Mathematics of Life

I was concerned with Darwin's rejection of mathematics and beginning to understand why my college biology professor was concerned over my interest in mathematics. Even though nearly a century had passed since Darwin's rejection of mathematics in biology, there were still many who agreed with him. More importantly, if mathematics was not useful for studying biology, what of the calculations I still supposed were behind President Truman's decision to bomb civilians?

I was relieved when I began finding others who had disagreed with Darwin. For example, the mathematician Karl Pearson (1857–1936) stated in 1901: "I believe the day must come when the biologist will—without being a mathematician—not hesitate to use mathematical analysis when he requires it."

Growth and Form: 1965–1967

A Scottish biologist, mathematician, and classics scholar, D'Arcy Thompson (1860–1948) described how to use mathematics in biology in his book *Growth and Form*. First published in 1917, the slim volume had grown to more than 1,000 pages as he revised it in 1942.[1] He wrote then, "My sole purpose is to correlate with mathematical statement and physical law certain of the simpler outward phenomena of organic growth and structure or form."[2] I grew increasingly fascinated as Thompson described the spirals of the chambered nautilus and the patterns in the leaves of plants with the same equation based on the Fibonacci sequence. This same series could also be used to describe the growth of a population of rabbits, at least simple rabbits. This was like Euclid's or Einstein's equations that described mathematical or physical features of the world, but Thompson's equations described biological features.

One of the features that he described mathematically was the growth in the size of individual people from birth to death. He used French census data to introduce concepts such as changing rates of individual growth and the difference between the growth of individuals and average growth of the population as a whole. But Thompson showed how similar mathematics could describe the growth of many types of organisms: microscopic algae (*Spirogyra*), crocus from the garden, giant tortoises from the Galapagos, cod from the Firth of Forth, and blue whales. He even described how the age of cod could be approximated from analysis of repeated measurements of the size of the fish caught over a few years. Curiously this involved looking for "baby booms," like the post-WWII glut of college students that had kept me out of the University of Oregon.

Importantly, in his second edition, Thompson included newer examples about the growth of the number of individuals in a group or population. He focused on one population in particular: humans. The Reverend Thomas Malthus (1766–1834), an English economist and demographer, estimated that human populations tended to double in numbers every 25 years and famously claimed in 1798 that with this compounding growth, the human population would outstrip growth its ability to produce food,

---

1. Thompson, *Growth and Form*.

2. Gould quotes several laudatory statements about Thompson's book, including Peter Medawar: "the finest work of literature in all the annals of science that have been recorded in the English tongue." Thompson's description of the mathematical beauty of nature was influential for many scientists and mathematicians and even artists. Gould, "D'Arcy Thompson."

the so-called Malthusian dilemma. Little did I know that my career would be tightly linked to Thompson's book and Malthus's dilemma.

Herbert Spenser's Definition of Life

One evening I got up from my favorite library carrel and wandered along the shelves of old books in a reading stupor. In the gloom, I happened to see a name I recognized from grade school, from the discussions of the Boiler Room Boys: Herbert Spencer. The book was one of a series titled *The Synthetic Philosophy of Herbert Spencer*. I wiped the dust from the cover of the first volume, *The Principles of Biology*.[3] It was published in 1865, just a few years after Darwin's *Origin of Species*.[4] In Spencer's preface, I read: "The aim of this work is to set forth the general truths of biology as illustrated by and as illustrative of, the laws of Evolution."

I was instantly hooked, just as I had been by Thompson's book on mathematical biology. Spencer wrote about general biological truths from philosophy but also from mathematics. His book seemed like a way to answer questions, to find truth. So, I began reading Spencer at the beginning, where he began by trying to define life.

I had experience with many kinds of life by the time I got to college. I recalled my tears over the rabbit whose life I took to get its skull as if the essence of life was in its head. I remembered my shame at my deceit in manufacturing a science fair experiment using tomato plants that I manipulated through hormones. I remembered making slime molds come together to form a fruiting stalk. In the far corner of our farm, I kept several hives of honey bees; capturing new swarms, burning hives that became diseased, and harvesting sometimes too much honey. I knew how to manipulate life, but Spencer's book showed me that I didn't really know what life was.

For example, he worked step-by-step through a series of definitions of life that he had encountered, identifying the limitations of each before moving onto another. After 73 pages he finally arrived at the best he could do. For him, something was alive if it was a "definite combination of heterogeneous changes, both simultaneous and successive, *in correspondence with external co-existence and sequences.*" That is, life involved making internal changes in response to external changes.

---

3. Spencer, *Principles of Biology*.
4. Darwin, *Origin of Species*.

That life could be seen this abstractly surprised me. He seemed to be saying that there were different things inside a living organism that were changing, like digestion and growth. Those things were connected to external things, such as eating food. The essence of life is that the food was digested into chemicals and energy and those then into growth.

Maybe this kind of definition would be sufficient for plants, like my science fair tomato plants, but it missed something that I had seen when I was collecting lizards. The professors dumping the leftover lizards instead of releasing them had brought up feelings that embarrassed me. But especially this definition did not describe what disappeared from the eyes of the rabbit as it bled to death.

I didn't know how to reconcile my experiences with Spencer's definition, and his style was becoming verbose and slow-going. I didn't have time then for a close study of the many volumes of Spencer. I read enough, however, to realize that he had taken on the task of answering many of the questions that my brother and I asked as children, so I tucked Spencer away in the back of my mind, certain that he could help me later when I had more time.

Aldous Huxley's Perennial Philosophy

The philosopher Aldous Huxley (1894–1963) was a cousin of the biologist Julian Huxley, mentioned above relative to Darwin's evolution. Both Huxleys were grandchildren of the biologist Thomas Huxley (1825–1895), a defender of Darwinism known as "Darwin's bulldog." Aldous Huxley described his quest to understand the effects of a derivative of a cactus plant called mescaline in his 1953 book *The Doors of Perception*. He was interested in the chemical similarities to adrenaline and the reports that mescaline possibly caused symptoms of schizophrenia. More than questions of the cause of mental illness, however, Huxley was interested in understanding more about the mental experiences of others. Thus he wrote, "I was convinced in advance that the drug would admit me ... into the kind of inner world described by Blake." He was referring to the poet William Blake (1757–1827), who wrote in his poem "Heaven and Hell": "If the doors of perception were cleansed, everything would appear to man as it is, infinite."

Huxley was disappointed in the effects of mescaline, however, and I was thereby discouraged from doing my own experiments, even if I could have. But I stumbled on another of Aldous Huxley's books, *The Perennial*

Philosophy, written in 1944. Aldous Huxley drew together his observations of the writings from many religions, immediately referring to both the Bible, which said, "Blessed are the pure in heart, for they shall see God," and a Sufi poet, who said, "The astrolabe of the mysteries of God is love." The wide variety of religious experience all seemed to be one, all tied together through love, and all pointing toward "the one divine Reality substantial to the manifold world of things and lives and minds."

I was taken with the conclusions that he drew from these many observations, especially that all religions point toward the one and same God. And I stopped just there, for he also said that God, the "Mind at Large," could only be understood through a systematic program of meditative experience. Based on such experience, he claimed that we could work out "a system of empirical theology," and thereby understand God. I could see that such experiences would require a lot of time, as he assured his reader, and his book was long. I stored away what I thought was his message, that there was a Mind at Large, and put the book back on the library shelf for another time.

Flavius Josephus's Jesus

"He is going home to North Dakota now. I don't know if he wanted to be successful," she added, watching my eyes.

I was still dumbfounded and couldn't respond. I didn't understand. I had never known anyone who would do this.

"He wasn't very good at it," she half-joked, betraying her Scandinavian dourness.

I had seen her walking my previous roommate across the university commons, his wrists obviously bandaged. I had started to cut across the grass toward that them, but she caught my eye and shook her head ever so slightly. I continued toward the library where she later found me in my favorite study carrel. She explained that he had slashed his wrists the evening before, but she had found him in time.

"His father was flying in to take him home, and I didn't think him seeing you would help just then."

"I intended to look him up this fall," I said lamely. "But I guess I hadn't tried very hard. He spooked me, you know."

"Yes, sometimes me too. But I think last year he was looking for your approval, do you know that?" she asked.

I started to reply "No," but my word came out a hoarse "Yes," not quite a sob. At some level, I knew what she was saying and was ashamed.

"He spoke of you as if you were Jesus," she continued. When you didn't seek him out this year, he was devastated."

She left me in my carrel, hidden away and full of regret.

That evening, still wondering about her words, I wandered down the library's shelves of old journals and books and saw the title *The Antiquities of the Jews* on the spine of a large volume by Flavius Josephus (37–100). A bell rang. Somewhere I had heard that a Jew named Josephus wrote about Jesus not long after his supposed death and resurrection. As I focused more and more on biology and mathematics, I had increasingly believed that Jesus was mythical. But now I had in my hand something that had been written about him over 1,800 years ago. There were 20 volumes of Josephus; how could I ever check to see if Jesus really was mentioned?

I returned the next night with more information: the campus pastor was familiar with Josephus and told me where to look. Blowing the dust from the cover of volume 18, I fumbled to the third chapter, and there in the third paragraph I read, "Now there was about this time Jesus, a wise man, if it be lawful to call him a man." I remember feeling a chill from the possibility that Jesus had been a real person, not just character in yet another Bible story. I read on as Josephus seemed to echo Missy Lane's Bible stories. The words sounded too much like they were written by a believer, a Christian, but Josephus wasn't. Were these words authentic, or had they been Christianized subsequently? I was jolted by the last line of the paragraph: "And the tribe of Christians, so named from him, are not extinct at this day." It sounded so anticipatory, like Josephus meant they were not *yet* extinct, but that he expected them to die out.

Of course, Christians are still not extinct today, although Jesus had not appeared on my philosophy charts from grade school. Spencer's and Aldous Huxley's philosophies were very different from each other, and Jesus' words in the Gospels were even more different. But they all addressed big questions, the kind my brother Eric and I had talked about. He and I had not answered these questions then, and now there were many more of them.

I returned to the philosophy folders from grade school, now on my dormitory room walls, and found the word that I was remembering, *Weltanschauung*. This harsh-sounding word is easier to pronounce in English: *worldview*. This was a way of approaching our big questions that was described by the German philosopher Wilhelm Dilthey (1833–1911). He had been inspired by Immanuel Kant's (1724–1804) argument in his eighteenth-century *Critique of Pure Reason* that objective and certain knowledge was possible, and he was seeking "an experiential science of spiritual phenomena," especially "laws governing social, intellectual and moral phenomena." This focus on laws, like Kant's objective knowledge, appealed to me. Dilthey, however, thought that worldviews based on natural law were too restrictive. There was much more of the human experience that we needed to account for. Dilthey's idea had been picked up by Abraham Kuyper (1837–1920), a Dutch Calvinist, who argued that "all knowledge proceeds from faith of whatever kind."[5] It is this faith that underlays the modern sense of worldviews, which can be defined as the fundamental beliefs people hold. Although a Christian, Kuyper realized that people other than Christians formed worldviews that they had faith in as well.

As I continued to wonder at the calculations about killing Japanese civilians in World War II and, more recently, why I felt so badly about failing my roommate, the idea of worldviews seemed promising. I tried to list some of the fundamental beliefs of Thompson, Spencer, Aldous Huxley, and Jesus. This quickly became overwhelming because everyone used the words differently; eventually, I put my lists aside. Thompson's mathematical approach to biology seemed like a secure route to knowing truth, while Josephus's Jesus reminded me of the uncertainties of Sunday school. I had chosen science, of course, and remained unconvinced about Jesus, so I suppose I was somewhere between being an agnostic and an atheist. But sometimes I couldn't help feeling nagged by the possibility that just maybe Missy Lane had been right about Jesus all along.

---

5. Bratt, *Abraham Kuyper*.

4

Breaking Down Belief

My first two years of college had introduced me to more questions than answers, especially questions about the church's answers to the big questions of life that my brother Eric and I had struggled with. I had succeeded in ignoring those problems, and especially the role of Jesus in all of this, for many years. Later my attention was drawn back to the concept of worldviews by a pastor whom I had been complaining to about my childhood Sunday school and the confusion between science and religion. I was acting like I understood the problem, that I had come to grips with it. After all, I was a scientist and should understand such things. When I responded somewhat vaguely to his question, "Have you read about worldviews?" he reached out to his bookshelf and handed me his copy of James Sire's book *The Universe Next Door*. "Read it," he said, a little curtly. The book began with two these two short verses:

> A man said to the universe:
> "Sir, I exist."
> "However," replied the universe,
> "The fact has not created in me
> A sense of obligation."
> —Stephen Crane, "War Is Kind," 1899[1]

> When I consider your heavens,
> the work of your fingers,
> the moon and the stars,

---

1. Crane, *Prose and Poetry*.

which you have set in place,
what is man that you are mindful of him,
the son of man that you care for him?
You made him a little lower than the heavenly beings
and crowned him with glory and honor.
—King David, Ps 8:3, circa 1000 BC

Those passages contrasted the worldviews of a Hebrew theist 3,000 years ago and an American nihilist 100 years ago. In contrasting such ideas, James Sire (1933–2018) was pointing out the differences in underlying beliefs, and his book effectively revealed to me the presuppositions, the beliefs, inherent in different answers to fundamental questions. I was quickly back to my yet unanswered childhood questions.

Sire was doing what I had tried to do in the university library when I first encountered worldview thinking, listing the underlying beliefs of different philosophers, but he developed these ideas much further. While I had long ago assumed that all the philosophies on the Boiler Room Boys' charts were true, Sire now showed me how to boil down a worldview into its parts by describing the answers to eight basic questions and by examining people's practical views of the world, views on which we base our lives.

The book proceeded systematically, beginning with his definition of what he meant by worldview: "A commitment, a fundamental orientation of the heart." I thought of my questions in grade school about the value of Jesus, Jews, and Japanese, in high school about honesty in science, and in college about the adequacy of science and my responsibility to my roommate. These were important questions, and they did engage my heart.

Worldviews can be expressed in two ways, as a story and alternately as a set of presuppositions. I wasn't sure about a story, but a list of suppositions that we hold about "the basic constitution of reality" sounded good; I could deal with a list. So I persisted with the book, buying his latest revised version to avoid marking up my pastor's copy of Sire's earlier edition. Sire listed a variety of possible answers to questions about the definitions of "prime reality" and "external reality," making a subtle but crucial distinction. He described ideas about the nature of humans and human death, central ideas that we rely on daily. He also spelled out ideas about where knowledge and ethics come from, and finally about the nature of history. The philosophical haze that I had encountered when I studied individual philosophers began to clear, and the attractiveness of comprehensive systems, such as Aldous Huxley's "Perennial Philosophy" began to fade.

But Sire's worldview analysis was more than just studying lists of possible answers to a series of questions to find correct answers. These were basic questions because there did not seem to be a way to prove the answers to any of them to be true. Instead, one had to choose answers that seemed right, answers that one could believe were true. To the degree we believed a set of answers to these basic questions, we made what Sire called a "core commitment" to a worldview.

Faith vs. Belief

Sire's description of worldviews included the many presuppositions that could be conveyed through a list or a story. Our belief in one or the other of the several worldviews based on those presuppositions provides "the foundation on which we live and move and have our being."[2] Such a foundation is not based only on an intellectual assessment of the likelihood of each of the several presuppositions being true. It is also based on our faith.

When Niagara Falls was first crossed on a high wire in 1859, the funambulist, or rope walker, was Jean Francois Gravelet, recently arrived from France. When he first stretched his two-inch hemp rope across the gap between the U.S. and Canada, most people who were able to inspect the rope stated that they did not believe he could do it. At 5 p.m. that day he picked up his 26-foot-long balancing pole, stepped off the United States onto his rope, and a few minutes later stepped off his rope and onto Canadian soil. Gravelet believed he could do it, and he made believers out of the thousands who came to see him repeat this feat over and over again.

Once, after pushing a wheelbarrow across Niagara Falls on his rope, he asked people in the audience if they believed he could cross back over Niagara Falls with a person in his wheelbarrow? Although many people expressed their belief that he could do that, when he asked for a volunteer to get in his wheelbarrow, no one stepped forward. Believing that Gravelet could walk the rope was easy; they had seen him do it repeatedly. Stepping into his wheelbarrow, however, required something more than belief. It required faith in Gravelet himself. Belief and faith to act are not the same things.

It is faith that is required, and everyone who embraces any worldview does so by thoughtful belief in the presuppositions. But to fully embrace it,

---

2. Sire, *Universe Next Door*, 20.

one must have faith to get in whatever wheelbarrow is on offer. We must be willing to act on our beliefs.

People put their faith in the beliefs that constitute their worldviews, whether or not they have thought through them in terms of Sire's basic questions. We live our lives based on that faith, getting in Gravelet's wheelbarrow again and again.

Once, on the last leg of a trip home, I sat next to a young woman on her way to one of the seven sacred mountains of the world.

"So, Mount Shasta is a sacred..." I fumbled.

"Mountain," she supplied. "It is where we can more easily access, well, they say the spirit world. God, I guess."

Realizing I was stepping in deeper, I asked, "What is your religion?"

"Mount Everest is another, and then there is Inyan Kara in the Black Hills in Wyoming," she continued, ignoring my question.

"What makes them sacred?" I ventured, following her lead.

"Well, they are like portals or a vortex concentrating energy, which makes you feel good," she replied brightly, but without enthusiasm.

As my airplane acquaintance demonstrated to me, we make people uncomfortable when we address their worldviews. Indeed, to challenge someone's worldview can be an act of violence.

I continued trying to sort out that young woman's worldview, sifting through Sire's array of answers to basic questions. Closest I could get was New Age, but Sire included many variations in describing that worldview. There are many answers to Sire's eight questions, but unlike the various New Age beliefs, these answers group into a few combinations, a few coherent worldviews.

In comparing worldviews, I quickly learned some surprising things. For example, although I was aware that some of the Beatles had sought out Eastern pantheistic monism, I was surprised to learn that the version they brought back to the West was different. They abandoned the deep futility inherent in the Hindu and Buddhist ideas of karma and reincarnation in the circle of life. For the Beatles, it became a positive opportunity rather than a negative point of despair. Similarly, I had also been unaware that the transcendentalism of Walden and Thoreau, which intrigued me in high school, was related to Eastern mysticism.

I remember a long conversation with a historian about the futility of another worldview, postmodernism, in the academic community. I embarrassed myself and frustrated him with my ignorance. He was forced to speak from within that worldview to explain to me how wrong it was. I came away confused then, but it all fell into place when I got to Sire's summary of the postmodernist worldview. For example, Sire noted the postmodernist belief that "truth about reality is forever hidden from us. All we can do is tell stories."

Two Opposing Worldviews

What Sire helped me identify was my question of how we know anything, the source of knowledge. People have an innate desire to know why, and I pursued such questions for years without wondering how we could know an answer. Not everyone believes that we can obtain objective knowledge, some arguing that there is no such thing, and some arguing that real knowledge is beyond rational apprehension.

My main question—what is really *real* and how can we know it?—began at an early age. I remember when my brother Eric told me that atoms are 99 percent empty space. The protons, neutrons, and electrons in atoms are like planets in the solar system, solid but with large gaps between them. These gaps in the solar system are so large that the most distant planet, Pluto, was long thought to exist only based on its gravitational effect on the movement of other planets. It was not visually observed until 1930, after a year of searching by the astronomer Clyde Tombaugh (1906–1997). That there were similar gaps within atoms freaked me out for a while.

So, what is reality? Some claim that reality is only these atoms and all their gaps. That is all there is, and everything that happens is determined by them. Further, matter is eternal and uncaring. Stephen Crane's poem "War Is Kind" rightly describes the implication of this belief about matter: our existence has not created in the universe a sense of obligation to care about us.

Sire attempted to answer this question by distinguishing between prime reality and external reality. For those who believe that prime reality is that atoms are all there is, external reality tends to be a closed system ordered by cause and effect. This was modeled by Newton and later by Einstein. The closed model belief is that external reality can at least conceptually be described by a mathematical model, albeit an enormously complex one. This group Sire terms atheistic naturalists.

In contrast, others believe that reality is better described as an open system. This is usually connected to a belief that this open system was created from nothing by God and, while ordered by cause and effect, is also open to his intervention. For this group, the answer to my initial question is: *yes*, matter is real and not an illusion, and it was created by God, who is eternal, infinite, and personal. Sire calls this group Christian theists.

Atheistic naturalists and Christian theists both believe that we can obtain authentic knowledge but differ on where that comes from. Both beliefs speak of human mental faculties allowing us to interpret our observations of the world to obtain knowledge, but naturalism would limit us to that source of information and often limit us to the methods of science. Christians, however, go further in believing that we can also obtain knowledge through other methods, including a special revelation from God and internal revelation from within us.

Besides knowledge, these two worldviews also differ in their understanding of human death, ethics, and history. They both see knowledge through reason and observations, but Christian theism also sees knowledge through communication with God. Atheistic naturalism sees ethics as being based on human ideas, as opposed to God's ideas. History then becomes either a purposeless or a meaningful sequence of events, the latter fulfilling God's purposes for us. Finally, the core commitments of these two worldviews are either oriented toward one's self or toward God, although the specifics here will vary among worldviews and among people's unique interpretations of their worldviews.

These two conflicting worldviews, atheistic naturalism and Christian theism, resonated with me, as I saw that I had chosen a road toward Christianity as a child, turned toward atheistic naturalism as a young man, and then back toward theism and finally Christian theism. The answers to Sire's basic questions inherent in atheistic naturalism included that matter is all there is, that there are only three dimensions plus time, and that humans are machines, albeit complex ones, whose behavior could conceptually be predicted based on physical processes alone. The answers for Christian theism were less concrete: that there is a personal God creating things like the universe and people and operating in the same three dimensions and time, plus an additional spiritual dimension. Thus, God created us in his image, with cognitive faculties that supersede the physical.

Science vs. the Bible?

Of all the historical and scriptural documents and all of the descriptions of personal experiences with God, the Bible is the single most important source for understanding Western civilization. It has been and continues to be the most important book in any language, having had a print run of over six billion (that is, thousand million) copies. The roughly 750,000 words have been translated and reprinted in over 1,000 languages.[3] It is a collection of Hebrew, Aramaic, and Greek scriptures written over roughly 1,600 years. The text begins with the origin of the world but quickly settles in to describe events in the Middle East and southern Europe over several millennia. It has gaps; for example, the second part, the New Testament, begins 400 years after the end of the first part, the Old Testament. Further, unlike many sacred texts, the Bible's historical authenticity has been validated by comparison to many other extant historical documents and by archaeological and historical studies.

The roughly 40 authors wrote in many literary forms, including "narrative history, genealogies, chronicles, laws of all kinds, poetry of all kinds, proverbs, prophetic oracles, riddles, drama, biographical sketches, parables, letters, sermons, and apocalypses."[4] While translating the words is demanding enough, interpreting what the words meant then (exegesis) and further what they mean today (hermeneutics) is even more demanding. Interpreting the Bible is central to understanding the observations people have made, the scientific, philosophical, historical, and theological observations that have defined Western civilization.

The story begins with the origin of the universe and the creation of man. It continues through the origin of a particular people group, the Hebrews, and climaxes with the death and resurrection of one man who claimed to be the son of God. The story ends with letters circulated among churches that developed in the decades following Jesus' resurrection. The outworking of this strange story occupied Western civilization ever since. To understand the relationship between science and religion requires an understanding of the Bible as a whole. As C. S. Lewis cautions, "We must not use the Bible (our fathers too often did) as a sort of Encyclopedia out of which texts (isolated from their context and not read without attention

---

3. Statistic Brain Research Institute, "Bible Statistics."

4. Fee and Stuart describes the nature of the Bible at length. Fee and Stuart, *How to Read the Bible*.

to the whole nature and purport of the books in which they occur) can be taken for use as weapons."[5]

Further, we must not use the Bible as if it were a science textbook written to explain facts about the world we find ourselves in.[6] We draw on Moses' description in the first part of Genesis and wonder what his purpose was and whether we should expect him to accurately report things that no one was present to have observed. Nonetheless, he wrote directly, and we should expect that what he wrote would not be contradicted by what he could have observed directly or by what we can observe today. Further, we should expect that he does not contradict himself within what he wrote and that other parts of the Bible would not contradict one another. Interpreting the Bible, both the specifics of Genesis but also the many more observations of the world and of people, requires that we carefully consider the range of possible meanings of words used in a specific passage in the context of the entire text and other historical information.

Determinism

Once, when I was in grade school, my brother challenged me: "I think you are a robot."

"I am not," I replied vehemently. "Robots only do what they are told to do, and I know that I am not a robot because I can do whatever I want."

"How do you know that what you think you want to do isn't just what you are really told to do?" he continued.

I saw the logical problem immediately and burst into tears.

A basic tenet of atheistic naturalism is that we all are robots, that is, that everything we do could be, at least conceptually, predicted from physics. Rene Descartes (1596–1650), the father of modern philosophy, in 1637 posed this question in terms of automatons, that is, machines that respond to human interaction. Later the eighteenth-century philosopher Denis Diderot (1713–1784) reframed the question in terms of the possibility of finding a parrot that could answer all questions. In 1950 chemist and science fiction writer Isaac Asimov (1920–1992) suggested in his novel titled *I, Robot*[7] that robots were far more complex. In grade school, I was plagued by Eric's question, which he reiterated by reference to Asimov's book, and

---

5. Lewis, *Collected Letters*, letter dated November 8, 1952.
6. Collins, *Science and Faith*, ch. 4.
7. Asimov, *I, Robot*.

I took to watching myself to see if I could predict what I was about to do. I think I was mimicking Asimov's main character, robopsychologist Dr. Susan Calvin, who had early learned to calculate parameter values for the positronic brain of robots to ensure that its "responses to given stimuli could be accurately predicted."

I was frustrated by Eric's question then and ended up shouting at him: "I know I'm real, but I don't know about you!"

That was a claim based on what I felt, frustrated by the determinism I sensed in Asimov's story but could not logically counter. What I had missed was that Asimov's three laws of robotics, which required robots not to injure a human being, drew just the distinction I was claiming. In Asimov's world, being a robot and a human was logically inconsistent. That is, there is something else besides determinism involved.

In the early 1950s, Alan Turing, a pioneer computer scientist and ultramarathoner, was also becoming interested in what has come to be known as artificial intelligence. Can machines be made that possess true intelligence? And if so, how could you know? Answering Turing's question has itself become an ultramarathon, with no answer for six decades now, regardless of there being an annual competition for a $100,000 prize held since 1991. Not for want of trying and despite tremendous advances in computing machines, neither a serious candidate for this intelligent machine nor an effective test of such a machine has been developed. This spectacular failure, nonetheless, is an observation of our world that tells us something. If people were created in God's image, then our intelligence is a result of that creation, and our failure to create artificial intelligence would be expected, in the form of abductive logic, to be "a matter of course."

Death and the Meaning of Life

Atheistic naturalism sees human death as the extinction of personality and individuality. I remember wondering about this at my grandmother's funeral. As I observed her lifeless body, I was shaken, suddenly remembering the wristwatch that Grandma had recently given me and then the woolen stockings I was wearing that she had knit. She knit everyone in the family stockings, quietly in her corner chair.

Someone whispered to me that Grandma was just sleeping in her coffin. I wondered if she were dreaming and whether that was fun or scary. I liked her and hoped her dreams were fun. I hadn't thought then of her going through a gate when she died, and I hadn't thought that the path divided at that gate. Some were sure Grandma wasn't knitting socks anymore, that she didn't exist. Others evaluated her life and predicted if she were separated from or with God. I couldn't ask her in her coffin, and I wondered then at how these questions could be answered. After Grandma's funeral, I so hated to sleep. Sleep became, for me, an enormous waste of time, as there were so many things to do. But that wasn't the real reason.

"Remember, we are cold nothing," said Victor Stenger (1935–2014) in wrapping up a debate about theism at California Polytechnical University in 2009.[8] Stenger was an outspoken naturalist, writing, for example, *God: The Failed Hypothesis*.[9] While most naturalists are not so blunt, this is what their worldview implies. To verbalize this conclusion is to conflict with the underlying belief of many people that human life has some larger purpose. Most people have a feeling that they have a purpose, and indeed that the absence of a feeling of purpose is deadly. For them, this is a mark against the presuppositions of atheistic naturalists.

Aldous Huxley's spiritual wanderings are a good example of people reacting poorly to the idea that their lives have no meaning. So poorly in his case that he was willing to change his worldview completely. Most atheistic naturalists, however, aren't so honest, usually failing to spell out the implications of the meaninglessness that their beliefs suggest.

Huxley wrote that he did not want the world to have meaning so that he could be free from the sexual restraints of Christianity.[10] He "went through several worldview changes in his productive literary life,"[11] always looking for a logically consistent understanding of the meaning of life. But Huxley couldn't hold a belief of meaninglessness on so weak an argument as sexual restraints, and he developed another worldview, lamenting that the "the scientific picture of the world is a partial one."[12] Eventually, he adopted the religious and philosophical syncretism expressed in his *Perennial Philosophy*, where I met him during my college years.

My college encounter with *Doors of Perception* and *Perennial Philosophy* had seemed transitory then, but over time, I began to realize that I had

---

8. See "Ross Debates at Caltech" at http://www.asa3.org//ASA/newsletter/janfeb09/janfeb09email.htm for Stenger's statement.

9. Stenger, *God*.

10. A. Huxley, *Ends and Means*.

11. DeMar, *Philosophy of Meaninglessness*.

12. A. Huxley, *Ends and Means*.

adopted some of Aldous Huxley's ideas. I couldn't have verbalized it, but I came away from his books believing in the irrelevancy of the distinctions among religions. There was something out there, but many paths to finding it. But Huxley's commitments to truth and morality changed with his philosophical transformations, which left me uncomfortable. In contrast, Christian theists claim that truth and morality are invariable because God doesn't change. Huxley's explorations of alternate worldviews were unsettling, but not as unsettling as the season in my life when my own worldview began to break down.

5

Breaking Down: 1968–1972

"I HAVE BEEN IMPRESSED with your fieldwork this year," my biology professor said one April day during my sophomore year, just as we finished a two-hour class field trip. "You have a very sharp eye. Would you like to go to Eniwetok Atoll as part of my field team this summer?"

My heart jumped at his invitation. I had been hoping for being involved in a biology field project since high school when the college recruiter had mentioned the possibility.

"Yes, very much," I managed.

"Well, I would like to have you under one condition," he continued. "That is, you would have to be more cooperative with me, and with the team. Sometimes I can see you bristle when I give you directions."

My face flushed as I remembered the flash of anger when he asked for the little beetle I found during one classroom field trip, a rare lumpy-backed snail eater. I didn't know this creature even existed, but after he told me what to look for, the search image, I quickly found one where he hadn't. My anger must have been apparent because he stopped the team and patiently pointed out that fieldwork was a team effort and that he was the leader of this team. This beetle would make an important addition to the department's insect collection.

That summer I made my first airplane flight, a short hop from the airport nearest to Camp Creek, north to Portland, Oregon. I remember experiencing for the first time the breadth of the Willamette River Valley from the air. I was ecstatic escaping Camp Creek, never dreaming that such flights would eventually become old hat. We stayed at Hickam Air Force Base near Honolulu for a couple of days and finally arrived at the Eniwetok

Marine Biology Station aboard the once-a-week Red Tail flight. It seemed like the end of the earth, but my heart soared.

Eniwetok is a string of islands forming a coral atoll for which the U.S. had battled Japan in 1944. By 1948 the U.S. had evacuated the residents and began using the atoll for testing more nuclear bombs. In 1954 the laboratory was established to study the recovery of the islands from the explosions. We visited the sites of the test explosions aboard a WWII landing craft, carrying our Jeep and supplies. We saw the remnants of towers and reinforced concrete buildings. We swam with blue-eyed scallops along the beautiful coral reefs next to the black waters of the 10,000-foot drop-off formed by the vaporization of the island of Elugelab.

The expedition lasted six weeks, and I was careful not to cross my professor. I was a big help because, unlike my professor, I didn't get seasick on the tiny landing craft that we used to dredge for crabs in the shallow atoll. Everything seemed to be going fine until the last week, as we were packing up our collections and equipment. I had been making my own collection of beautiful shells and Japanese fishing floats that washed ashore there. When my professor saw those shells, he quietly reminded me that we were a team. Again, my anger flared up, but this time I could control it—barely.

Finishing College

A week after I returned from the South Pacific I married Margene as planned, and we settled into cozy isolation from the world. We rented an apartment at our university in Tacoma, Washington, one in a ramshackle row of apartments that had been quickly built for veterans returning from World War II, my father's generation. The rent was $40, and groceries at the Piggly Wiggly store averaged $5 a bag. We had a car now thanks to the stipend from the Eniwetok expedition and enough gas to get home to visit family in Oregon was also $5. I drove a school bus, which, along with occasional $50 checks from Mom and Pop, gave us enough to meet our needs but little more. The November 1967 student protests of the developing war in Vietnam were focused in Washington, D.C., but these passed over our heads virtually unnoticed.

Entertainment was always a problem, and the default was having neighbors over for drinks and a hand or two of cards. We experimented with many types of drinks and played many different card games. I liked the drinks but, after constantly losing to my father growing up, still did not like the card games. I always seemed to have trouble remembering the rules, and I didn't like the hiddenness. One night the German couple next door brought over ginger beer, and we were soon well into too many hands of cards and too many Moscow mules. I remember stumbling into the toilet and falling against the shower stall with a great noise. Margene rushed in to help me up, inadvertently stepping on my shoelace. I needed to pee but couldn't because she was holding my foot down, and we all started laughing hysterically. That passed for entertainment.

The next year our isolation was interrupted by the 1968 Tet Offensive in Vietnam, the assassination of Martin Luther King Jr., and the election of Richard Nixon. When Nixon was sworn in as president, I was trying to complete the usual rituals associated with college graduation, plus one not so usual. Protests were increasing against the continuing escalation of the war, and Margene and I began evaluating our options. My automatic student deferment from the military draft would expire when I finished college, and I had already been rejected by the U.S. Navy's Officer Candidate School because of the migraine headaches I had suffered since early in high school.

If I got into graduate school, I might get a new deferment, but the draft boards were under pressure to supply increasing numbers of foot soldiers, "cannon fodder" we joked darkly as if we knew. There were rumors of replacing the draft deferment system with a lottery based on birthdays, which would give us greater security if I didn't win it. Otherwise, we discussed emigrating to Canada, but it wasn't clear what our future status would be; draft dodging was not popular in either country. We weren't prepared to abandon our families, and I wasn't sure how Pop would feel about me avoiding the military draft that had captured his generation. In the end, we decided to ignore all these uncertainties and move forward with my dream of a PhD, hoping not to be drafted.

One graduation detail was qualifying for a major. I didn't have enough credits for a major in either the biology or the mathematics department, and my biology professor urged me to take a couple of laboratory classes to get the major. I knew that I needed more mathematics and that I hated laboratory classes. Eventually, I went around the biology department's rules by appealing to the administration to define a multidisciplinary major. I argued on the basis of Thompson's book *Growth and Form* and an advertisement I found for a new graduate school program in what was called biomathematics at the University of Washington in Seattle. Getting the multidisciplinary degree turned out to be more important than I expected.

The Interview

"Why do you want to be in our biomathematics graduate program?" the interviewing professor demanded with a sigh.

"Well," I stammered, "because my father could have been a scientist if World War II hadn't happened. But also," I rushed on realizing my mistake, "because I am really interested in how biology and mathematics fit together."

"I see that you were allowed a multidisciplinary major because of your interest in D'Arcy Thompson's book," said the interviewer, generously leading me from my embarrassment. "We are particularly interested in your application because of that."

"What part of our biomathematics program are you more interested in," he went on, "medical statistics or wildlife studies?"

"Definitely wildlife," I responded.

"Good," he offered. "That is my area of interest too. What did you find most interesting in Thompson's book?"

I had read portions of Thompson's book on my own, but neither my college biology nor mathematics departments had been interested enough to teach anything from it. I felt a little uncertain at my understanding and wondered how specific an answer to attempt.

"Well, I liked when he described growth mathematically," I offered.

"The growth of individuals or the growth of populations?" he asked, putting his finger on the specifics, exactly as I feared.

"Uh," I hesitated. "I guess I got a little confused there. Thompson kept shifting between individuals and populations and then showed data on the frequency of cod lengths. I think all of those topics were related, but I couldn't quite get it."

"Yes," he said. "Thompson is a little confusing there. And what did you think of his thing on the growth of blue whales and their calves?"

I knew by then that I was out of my depth and hoped he wouldn't go any deeper into that book. I almost regretted making it a big thing.

"I was amazed that blue whales got so big, over 90 feet as I recall," I said, trying to pull up from a steepening dive.

"Well," the professor said, responding to my lead, "I guess that is enough about that."

I was sure that he wasn't impressed by my knowledge of blue whales and afraid I wouldn't get in. I could visualize receiving a "Thank you, but no" letter is a week or two.

But he continued, "My area of research is populations of whales and seals. There is so much we don't know. If you are interested in whales, I would be happy to have you as a student. I'd like to offer you a scholarship in our program, covering tuition and a stipend for living expenses. Are you interested?"

The Physical

The letter from the draft board denying my application for a deferment had come in September 1969. I had begun graduate school classes the previous July just one month after getting my college diploma. Then on November 26, 1969, President Nixon, by then under intense political pressure from student protests, called for a lottery for the military draft. On December 1 in a well-publicized visual demonstration of fairness, Congressman Prinie drew all the capsules containing the numbers from 1 to 366 from a huge glass cylinder; he drew my birthday on his third try.

Soon afterward I received the second letter from the draft board, now asking me to come to a room in downtown Seattle for a physical examination. I found myself part of a slowly moving line of young men stripped to our underwear in a drafty old building in the Skid Road district in Seattle that the army had rented for the occasion. The poking and prodding continued, each of us holding our paperwork. Then toward the end, a question was asked.

"Yes," I responded in my turn, "I do have some questions for the doctor."

"Yes, sir," I responded when told to wait in a second, adjoining room.

The day had been creepy, and images of civilians in WWII concentration camps seemed to overlay this line of young men. I was caught up in something much bigger than me, and all day that something had been systematically consuming the young men in this line. I was the last one to be called from the second room.

There the grandfatherly looking doctor relaxed after he saw that there were no more young men waiting for him after me. He asked about my concerns and then picked up my file. He noted I was married and in graduate school.

"Any children?" he asked, looking up from my file.

"One soon, a girl," I answered, surprised he had asked. "We are due this next September."

I knew babies no longer disqualified men from the draft, and we had chosen to become pregnant because we had been delaying starting our family too long. However, I didn't try to explain this to him.

He looked down again, and his face showed surprise when he got to the letter from my pediatrician concerning medication I had been prescribed for migraine headaches when I was in high school. He was outraged that I had continued to use it for so long to reduce my symptoms.

"How often do you take that drug?" the doctor asked, seemingly gently.

Two weeks after that physical, I got the third letter, instructing me to go to a neurologist's office also in downtown Seattle. There I found myself lying flat on a gurney, lights pulsing and flashing as the glue holding the electrodes on my skin dried and pulled.

"I'm simulating a horseback ride down a tree-lined lane," the doctor explained. "The flickering shadows should bring on one of your migraines."

The pulsing lights varied over the next half hour, punctuated by more questions and mumblings from the doctor. He could not induce a migraine. I was disappointed. A few days I got the fourth and final letter from the draft board. The neurologist had summarized his findings, and the draft board assigned me a status of 4-F, unfit for military service. I had won again, but wondered why, when so many of my friends were being drafted?

The Birth

A few months after I was declared unfit for war in Vietnam, I drove through every red light going to Capitol Hill early one Sunday morning. Margene had woken nauseous and in pain. A quick call to Swedish Memorial Hospital and we were on our way. A doctor and a nurse were at the ER entrance when we pulled up 15 minutes later. Margene was having a seizure, and the doctor said that she had either toxemia or acute eclampsia.

"We'll have to sedate her until the seizures are under control, then a C-section," the doctor said. "We need to get the baby out as soon as possible. They are both in extreme danger."

I didn't know what to think as she was wheeled away from me through the blue-gray swinging doors.

"Just fill out this form, sir," the clerk at the desk repeated.

I couldn't think as the full weight of the doctor's words began to settle on me. Would Margene die? What about the baby? It was the eighth month; what had the insurance company said about denying our coverage if the baby is premature?

"It's funny the details you think of," I said to myself, surprised that my mind went to the insurance problem.

The clerk shook my arm.

"Sir, please. We need the forms filled out now so they can begin treatment. You must focus, now!" she said.

I was pacing the waiting room when the doctor came up behind me an hour later.

"Your wife is almost fully sedated now, but her kidneys are failing," he said. "We will begin kidney dialysis as soon as we finish the C-section, and we will begin the C-section as soon as an operating room becomes available."

We were definitely failing our Lamaze class.

Later a nurse reported, "Your baby is alive, but her prospects are poor. Her APGAR score was only one out of a possible ten." She explained what that meant, and I thought I should be afraid—but I wasn't.

"Your wife is still sedated," she continued, "but I see they have her scheduled for peritoneal dialysis as soon as she wakes up."

"When will I be able to see them?" I asked, still dry-eyed and calm.

Eventually, I stood outside the glassed-in room with several rows of incubators, some with other babies. Our baby was still blue, and after a while the nurse who would carry us through this, Smitty, let me stand next to her incubator as the hours dragged by. I left a message for Herb, our pastor at our Lutheran church, who had been away hiking in the Cascade Mountains, and asked him to come to baptize our baby, Rachel Leah. I had been attending Margene's churches since we married, not because I believed everything or anything that was being taught, but because there seemed to be *something* there, and besides, our home life was better when we were attending church together. An emergency baptism sounded strange to me, but I knew that Margene would want this, just in case.

Once in a while, I would catch a glimpse of something to my left, always just out of my field of view, like a floater in my left eye. That was eerie, but at the same time reassuring, like I wasn't in this alone as my wife and our baby fought for their lives.

When Pastor Herb finally arrived that evening, Rachel had begun to pink up, but the oxygen tube was still taped to her nose. I left the incubator room to greet him.

"The doctor says the baby may not make it," I told Herb calmly.

"I understand," Herb said. "Margene would want this." Smitty gave him a mask, gloves, a cotton swab, and a bottle of sterilized water.

"What is her name?" Herb asked.

We went into the incubator room, and Smitty opened the access door to Rachel. Baptism doesn't take long for Lutherans.

As we sat in the waiting room later, Herb looked at me and said, "You are so calm. Why?"

"I think it's because I kept seeing something, or someone, on my left side, always just at the edge of my vision," I said.

Herb nodded but didn't speak.

"Do you know what that was?" I asked.

"I think it was Jesus," he answered quietly. "He tends to make people calm."

I didn't know what to say to Herb. I believed that Jesus had lived, remembering my college encounter with Josephus. But I didn't really believe that he was alive today. I wondered about my unbelief after both Margene and Rachel survived that day.

Leukemia or Elephants?

In the biomathematics program, students took biology and mathematics classes for two years, then sat for a set of four qualifying exams, and finally researched a dissertation project. Classes were easy, but my first setback was when I failed the first exam. We had put everything into this PhD; it was our pearl of great price, as Pastor Herb had pointed out.

I did pass that exam on my second try, and then the other three in short order. Then I had to choose a dissertation project. I had to present something to a committee and didn't have a clue what to do. I resented the committee's control and rented a canoe in the arboretum. As I paddled, I continued to think about quitting. My friends had counseled me that that would be stupid, that I was so close. But I felt the committee's control in my gut and decided to get a job. When my applications for teaching positions at community colleges were almost all ignored, I gave that up. Instead, I pushed aside my conflicts with my committee and set to work interviewing professors to find a project.

One professor encouraged me to study methods of doing a statistical analysis of leukemia drug trials. Unlike wildlife, where he said the data is always bad, he assured me that I could design proper scientific experiments and would soon have good statistical data. My brother Eric at that time was losing his long battle with leukemia. My hope rising, I asked about the likely success of the planned treatments.

"Well," the professor confessed, "it's not about cures, more about how different drugs might extend patients' lives, perhaps by months or a few years."

I flashed back on my manufactured science fair experiment; the control and treatment groups of tomatoes had both ended up in the garden. In the leukemia trials, what would happen to people who were assigned the placebo if, in fact, the treatment were successful? Who was going to explain that to them or to their families? Just as in my science fair experiment, my professor's answers to these questions did not convince me.

"It would not be up to us to decide those questions. You'd just be analyzing the results of the experiments," he explained. "Besides, the people agreed to be a part of the trials and know the risk of receiving the placebo." Again, like in high school, something I couldn't identify wasn't right with his answer.

I couldn't face those questions, especially with my brother in the hospital, so instead, I began looking for a wildlife project. A good friend and I had published a class project, a computer model of harvesting populations of African elephants. He encouraged me to follow that up. I did have some ideas about how I might extend the mathematics of our model, but they seemed pretty weak to me. I discussed those ideas with my major professor, and he surprisingly encouraged me to present those ideas to my committee. I would have to give a 20-minute description of what I was thinking, something that left me terrified. But they liked the new model and suggested that I apply it to a couple of examples.

Then they asked me to step out onto the third-floor balcony so they could vote formally on my proposal. I felt reassured by the committee, but time dragged by on the balcony, first 20 minutes, then 30, then nearly an hour. My anger against the committee started to resurface, and, once again, I began to make plans to leave school without my prized PhD. Suddenly the door swung open, and my major professor was standing there, looking a little frantic.

"I'm so sorry," he blurted out. "We finished our vote and our recommendations and then left and forgot about you out here. You won our approval, and we are looking forward to seeing your dissertation as soon as possible."

Then, noting the expression on my face, he joked wanly, "I'm so glad you didn't jump."

"They Shoot Horses, Don't They?"

My major professor had joked to cover his error, but it struck deep. Not everyone wins in that situation, remembering my freshman roommate who slashed his wrists. A few months later Margene and I went to the movies with Pastor Herb and his Scandinavian wife, Berit. Toward the end of the movie, the main characters realize they had won the Depression-era dance contest but that there was no prize money left after their expenses were deducted. In despair, Jane Fonda's character, Gloria Beatty, gives a pistol to her dance marathon partner Robert, played by Michael Sarrazin, and asks him to shoot her in the face. Sensing his incredulity, she justifies her request by repeating the movie's title, "They shoot horses, don't they?" Robert pulls the trigger and during his murder trial recalls that in her next moments, "She was relaxed and comfortable, and she was smiling. It was the first time I had ever seen her smile."

Afterward, we sat in a small North Seattle coffee shop. I was stunned at Gloria's question, still ringing in my head and my heart, and Robert's answer.

"How could Robert do that?" I asked.

"Nihilism," Berit tersely summarized. "We Scandinavians seem to have a corner on it." When she saw I didn't understand, she continued, "It's when people think that they have no intrinsic value. It's complicated, but nihilists believe that nothing has value, sometimes that nothing is real. We just are, and then we aren't. Sometimes it's less painful just not to be."

"People really think that way?" I finally asked.

"Especially when they don't feel their life has meaning," she replied.

I remembered my roommate's attempted suicide, and the questions his girlfriend had left with me flooded back. I did know that my roommate had wanted my approval and that in denying him that I had denied him a sense of meaning, of value. I understood now through her gentle words that I had failed him, and tears sprung to my eyes.

A Successful Suicide

Late one autumn afternoon, a Friday, I carried the draft of my dissertation up the worn gray steps to the dilapidated post-WWII office building. I intended to drop the draft off with my major professor's secretary for his review, but she wasn't in. Not wanting to leave it in the chaos of her desk, I pushed open the door to his darkened office.

"Yes?" a voice asked quietly, surprising me.

I blushed at being caught barging in and stammered something. Without turning toward me, he spoke in an uncharacteristically soft voice, too soft.

"It's okay. Just leave it there on the corner. I'll look at it this weekend."

But he would never look.

The next morning Tom, my current employer at the university, called to tell me that the police had found my major professor in his cabin on an island in Puget Sound, his .45 caliber pistol still in his hand.

I tried to ask a question but failed to get it out. Then Tom tried to answer what I couldn't ask but failed. We talked incoherently for a while.

Finally, Tom asked, "We can't easily cancel our trip to Fletcher's on Monday. Would you still be willing to go?"

I mumbled that I'd have to get back to him and hung up. I started working for Tom a few months ago, and this research expedition to study the water under an iceberg in the Arctic Ocean had long been scheduled for the last flight of the daylight season. We were to fly a commercial airplane to Fairbanks, a military transport to Barrow, and then an enormous Herc, a C-130 Hercules turboprop military transport, to Fletcher's Ice Island now floating near the North Pole. Tom was studying life under the ice, sampling animals and plankton through a two-meter diameter hole his team had drilled through the iceberg and continued to keep from refreezing.

That evening was yet another church potluck. I was numb and quickly retreated from the noisy basement to the sanctuary. I had never felt the meaning of *sanctuary* so acutely. I wept and wondered at my own failure to have seen the pain that my professor must have been in. I couldn't imagine enough pain that would lead anyone to turn towards the darkness of death.

After a while, Pastor Herb came in, and we talked for a long time, first about my major professor and then about the movie *They Shoot Horses, Don't They?* I remembered my college roommate, his girlfriend's gentle accusation. I couldn't simultaneously take in the movie, my failure to engage my roommate, and now my failure to engage my professor when I had heard his too-quiet voice.

"Herb, I feel like I failed my professor. Should I have said something?" I asked.

"I don't know, Tim, because you didn't know that he would kill himself. When someone threatens suicide, we should always respond, but sometimes we just don't know enough."

"I knew there were problems for my freshman roommate, and I didn't engage him then. And I knew something had changed with my professor, and I didn't speak to him on Friday, and now this," I said, tears starting to flow.

"Even when you can see signs later," Herb replied, "you don't know what they are planning. It's more important to engage with people when you see changes than to always be on suicide watch. That's God's job, to bring people who care to speak into people's lives, so they don't feel alone."

"Could I have helped by saying something?" I asked.

"That's a question I've asked myself many times, too many times," Herb said. "It is an impossible question, a question that brings more pain than light. People who kill themselves aren't ending the pain, but just passing it on. You have some of his pain now, and there is no way to go back in time to avoid that."

Herb and I discussed my plans about going to Fletcher's Ice Island in the morning. Everything seemed so up in the air, and I couldn't choose between staying for the funeral and going with Tom to the Arctic Ocean. Herb had made sense, that there was no way to go back. What I chose now should be about what I needed because there was no way to help my professor now. I thought I needed to be away from this death.

We left from Sea-Tac International Airport early the next morning. The flight from Barrow would be longer than planned because Fletcher's Ice Island had drifted very near the North Pole now. I was going as far as anyone could in order to get away from the darkness of suicide; that was what I thought I needed. As the giant Hercules C-130 descended towards Fletcher's Ice Island, I watched the whiteness stretch out toward the remnant of the sun setting over a ridge of ice. The streaks of sunlight were giving way to the months of darkness that I knew would come to the top of the world. The Arctic's darkness was not the darkness of suicide but was darkness just the same; I had to face it. At least here, looking up into the void, there was still a glimmer of hope among the stars.

6

Facing the Void

QUESTIONS ABOUT THE ORIGIN of the universe had resulted in major conflicts between my church and my science classes. My church's insistence on a recent origin of the universe and the Boiler Room Boys' explorations were inconsistent. The scientific claim that the universe had no beginning seemed so much simpler but what I hadn't known then was that the scientific claims had not been tested by actual observations. Instead, they came from the current unquestioned touchstones, or presuppositions, that were widely held. My own touchstones began to change by the time I started to examine the current scientific and religious answers to the question of the origin of the universe.

A Beginning?

On the religious side, I read the Judeo-Christian scriptures about creation. Moses, traditionally thought to have written Genesis somewhere around 1500 to 1200 BC, and St. John, one of Jesus' disciples writing around AD 84, used the same words in asserting that the universe had a beginning. The first two words of Genesis, בְּרֵאשִׁ֖ית בָּרָ֣א, translated from the right to left in Hebrew, and St. John's first two words, ἐν ἀρχῇ, translated left to right in Greek, both mean "In the beginning." The scriptural basis for believing that the universe had a beginning is clear.

On the science side, I began to read of Einstein's work and was surprised at his belief that the universe did not have a beginning. He demonstrated the strength of his faith in this by modifying his 1915 publication on a theory of gravity when it was pointed out that as published his theory implied that the universe did have a beginning. That is, he climbed into Gravelet's wheelbarrow of faith and inserted a fudge factor, the so-called cosmological constant, into his theory so that it now implied a constant universe. Adding this factor "was not justified by our actual knowledge of gravitation," he noted, but it was required to support his faith that the universe had no beginning, contrary to Moses and St. John.

Einstein's cosmological constant appeared in my junior high school science book, and I believed Einstein. However, when I was in high school, Einstein fell out of Gravelet's wheelbarrow. Cosmology had been slow to let go of the idea of an unchanging universe, despite Hubble's accelerating galaxies. But then in 1964 two American astronomers, Arno Penzias (1933–) and Robert Wilson (1936–), measured the temperature of what had been called the "radiation of the stars" to be -455 °F. They found that this radiation, called cosmic microwave background radiation, comes to the earth from every direction in the universe. It was soon argued, using abductive logic, that this observation would be "a matter of course" if the universe had a beginning. Einstein had, by then, revised his theory by eliminating his cosmological constant to again allow for such a beginning. Suddenly I realized that science could be mistaken and that in this case Moses and St. John had been correct all along. This reversal shook me but surprisingly did not deter me from science. Somehow, I ignored this massive scientific error and chose to believe that this was just part of the process by which science eventually would be able to explain everything.

But looking back now, that reversal of belief by cosmologists was a result of what Ludwig Wittgenstein (1889–1951) called the "modern system" when he wrote: "... in the modern system it should appear as though *everything* were explained."[1] Scientists like to reassure that their theory, in fact, explains everything, even if, in fact, it doesn't. To accomplish this appearance, they shy away from describing their uncertainties and assumptions, their presuppositions, and just claim they are correct. Perhaps an unchanging universe had been assumed because people have a low tolerance for the ambiguity inherent in science. Attempts to smooth over those uncertainties using assumptions result in frequent reversals of what is claimed to be true. Nobel Prize–winning physicist Max Planck (1858-1947) recognized these reversals as early as 1919, writing, "Science progresses funeral by funeral."

In fact, such reversals of what scientists claim is "scientific truth" have been so frequent that the philosopher-physicist Thomas Kuhn (1922–1996)

---

1. Wittgenstein, *Tractatus*, page 87, proposition 6.732.

in his 1962 book *The Structure of Scientific Revolutions* developed a theory about why they occur. Kuhn distinguished "normal progress" in science from what he called "paradigm shifts." The normal accumulation of observations confirming the current "truths" of science are held onto by scientific academies agreeing to ignore inconsistent observations, such as Hubble's galaxies. Eventually, however, enough inconsistent observations accumulate to force a change. In this case, the observation of cosmic microwave background radiation confirmed Hubble, and the balance tilted toward a new theory, pejoratively called the Big Bang. When this new theory emerged, it was quickly fleshed out and, in the manner of Wittgenstein's "modern system," cosmology would once again claim that everything is explained. Moses was right and cosmology wrong, but of course the new theory now explains everything, and again cosmology was made correct.

When Did the Universe Begin?

Job (9:8) and the prophets Isaiah (42:5) and Zechariah (12:1) all described the universe being stretched out, using the same word, נָטָה, transliterated *natah*, meaning "stretching." *Stretching* suggested that the universe did not begin full-sized, and this idea is consistent with the theories that cosmologists were pushing in the 1960s. Beginning instantaneously, which we glibly refer to as the Big Bang, was not like a nuclear bomb nor a cosmic piece of dynamite. Rather, the Big Bang shows up in cosmologists' mathematical models as a "singularity," that is, an impossibility due to dividing by zero. The interpretation is that there was an infinitely small space where all matter and energy were compressed.

The Big Bang is the phrase used to describe the release of that compression, the beginning of the universe, the beginning of all that exists in the material world, the beginning of time itself. This "stretching" is analogous to the rising of a loaf of raisin bread, with the galaxy-raisins gradually moving away as the loaf rises under the force of carbon dioxide gas released by the yeast. Cosmologists hypothesized that light eventually escaped from the dense soup of hydrogen atoms that were forming as the universe stretched, resulting in a uniform glow in every direction, cosmic microwave background radiation.

Further, cosmologists have developed and continue to improve models of the expanding universe based on interactions among subatomic particles and atoms. Those interactions are thought to be due to four fundamental forces. First, the building blocks of matter, known as quarks, are drawn together by the strong nuclear force. The subatomic particles are drawn together by electromagnetic forces, and radioactive atoms are taken apart by weak nuclear forces. Gravity at the macro level draws matter together. All this results in the formation of more complex atoms through nuclear reactions in stars.

Scripture doesn't say how long the universe has been stretching, but details of Hubble's observations of galaxies moving away gives us an idea about this. He showed that the galaxies are moving at different speeds, and the more distant galaxies are moving faster than those that are closer; that is, the galaxies have been accelerating. If the galaxies have been accelerating at a constant rate, we can mathematically compute the time since the universe began stretching. Simply divide the current speed (distance per second) by the acceleration rate (distance/second/second), the units canceling out to give an estimate of the number of seconds since the galaxies began moving, that is, roughly since time began. Hubble's estimate of the age of the universe was initially between 7 and 20 billion (that is, thousand million) years. Since then, with many additional observations of the speed of galaxies, this estimate has been refined to around 14 billion years, or more precisely 13.7 billion years. Independent of this, the systematic peaks and valleys in the temperature of the cosmic background radiation also suggest that the universe is roughly 14 billion years old.[2]

What Caused the Universe to Begin?

Although cosmologists had provided estimates of when the universe began, they and everyone else have obviously been unable to make any direct observations about the cause of that beginning. Lacking such observations, cosmologists have identified four primary forces of nature—the strong and weak nuclear forces, the electromagnetic force, and the gravitational force—and calculate from their mathematical models that the numerical values of these forces are necessarily very precise. This "surprising observation" implies that life would be impossible if any of those values were even slightly different. We don't know what determines the strength of these forces. That they were set at the time of the Big Bang by chance is unlikely because the mathematical likelihood of them taking exactly those values is

---

2. Dalrymple, *Ancient Earth*, 194.

vanishingly small. But if the values of these forces were not set randomly, how were they set?

The precise values for the four forces of nature led the physicist Freeman Dyson (1923–) to conclude, "As we look out into the universe and identify the many accidents of physics and astronomy that have worked together to our benefit, it almost seems as if the universe must in some way have known we were coming."[3] Interesting phrases, "almost seems as if" and "had known we were coming." These phrases catch many cosmologists' reticence to embrace the apparent design of the universe.

In contrast, in the last part of the first sentence in Genesis that begins this chapter, Moses was clear when he wrote אֱלֹהִים אֵת, transliterated as *elohiym bara'* and meaning "*elohiym* created." Leaving aside, for now, the meaning of *elohiym*, Moses meant that the universe was created and did not just spontaneously appear of its own accord. This is consistent with the so-called Kalam argument, an idea originating with medieval Muslim philosophers, that a beginner is logically necessary for anything that has a beginning.[4]

The Kalam argument and the precision of the forces of nature both suggest that a beginner began the universe. Moses' beginner was *elohyim*, the Hebrew word that accounts for 78 percent of the Hebrew references to God in the Bible. Moses describes some characteristics of God. For example, in the same passage, he states that God has a רוּחַ, transliterated *ruwach* and meaning "spirit." Scripture also describes many other aspects related to the creation of the universe. St. John identified the existence "from the beginning" of *logos* (the Word) and *theos* (God) (John 1:1). He went on in verse 3 to clarify that God created from nothing, not from something: "All things were made through him, and without him was not any thing made that was made." Similarly, Solomon's anthropomorphic wisdom, חָכְמָה, transliterated *chokmah*, was a characteristic of God from before creation. Wisdom speaks (Prov 8:22–23):

> The Lord possessed me at the beginning of his work, the first of his acts of old.
>
> Ages ago I was set up, at the first, before the beginning of the earth.

Moses went on to describe the object that Elohyim created as *shamayim 'eth 'erets*, or the heavens and earth, that is, everything. It's not that there

---

3. Dyson, *Disturbing the Universe*.
4. Craig, *Reasonable Faith*.

was preexisting "stuff" that God used to create everything else, that is, compressed stuff exploding out of the cosmologists' singularity. Rather, the universe was created de novo, from nothing.

Genesis is not the only account of the creation of the universe. For example, some accounts involved construction metaphors. David described Elohyim assembling beams whilst riding a chariot of clouds on "the wings of the wind" (Ps 104:3). Alternately, Chinese texts from as early as the sixth century AD describe the offspring of Ying and Yang, *P'an Ku*, chiseling the universe from chaos over some 18,000 years, after which he died, apparently so that his creation might live.[5] There are many more such texts to choose from, but we don't have much basis for choosing among or interpreting these ideas concerning the nature of a beginner beyond the apparent reliability of the source texts.

Some would like to avoid the force of the Kalam argument and the force of Dyson's observation of design by examining in more detail just how the subatomic world works. Quantum theory is a mathematical model that allows creating nuclear explosions, and one might think would provide insight into how the universe actually works. For example, perhaps there has been more than one universe. Suppose our universe's formation through the Big Bang is merely one episode in a repeating series of universes, cyclically expanding, collapsing, and expanding again. If at each hypothesized successive Big Bang the values for the fundamental forces of nature were randomly formed, then with enough repetitions a workable set of values might have occurred. The fact that we are here to consider this question is evidence that at least one Big Bang resulted in a universe that could support life, no matter how unlikely this might have been. A variation on this idea that doesn't require cycling is that perhaps an infinite number of universes were created simultaneously and existed in parallel. Either way, no beginner needed. There are many more ideas based on quantum theory about how the universe might have begun that don't require a beginner.

Physicists disagree, however, on the potential of such explanations, some arguing that such multiple universes are inherent in string-theory-based-cosmological theories, but others claim that this is just untestable speculation. For example, theoretical physicist Peter Woit (1957–) argued that "This line of inquiry has become a concerted effort to build a theoretical framework perfectly insulated from testability and sell it to the rest of the

---

5. The *Encyclopedia Mythica* includes many interesting creation myths. http://www.pantheon.org.

physics community and the public, hoping no one notices the circularity."[6] Clearly, not everyone agrees on the interpretations of quantum theory and how they might help decide among such theories.

How Does the Universe Work?

The atomic bombs that dominated my childhood were possible because of the theory of quantum physics for which Werner Heisenberg (1901–1976) was awarded the Nobel Prize in 1932. This theory was further developed by several physicists over the next decades, including Nils Bohr (1885–1962), Wolfgang Pauli (1900–1958), Max Born (1882–1970), and Einstein along with many others. The first nuclear explosion demonstrated the accuracy of the predictions from this theory. The Trinity explosion occurred in New Mexico in the Jornada del Muerto desert on July 16, 1945. The historical name for this area translates as "Journey of the Dead Man," and the Trinity explosion marked the first step in the "Journey of the Dead Man" that we have been on ever since. The uranium-based bomb named "Little Boy" dropped over Hiroshima was the next step, followed within days by the plutonium-based bomb named "Fat Man" dropped over Nagasaki. Although those were the only two atomic bombs used in war, this ongoing journey has dominated all life ever since. We should understand what we've learned from this journey.

While quantum theory predicted how to make atomic bombs, it also raised many questions about how the subatomic world actually works. The theory used to calculate bombs makes some interesting suggestions about how the universe works, suggestions that are counterintuitive and mystifying. For example, Heisenberg argued that everything behaves as both a particle and a wave simultaneously. Further, he demonstrated that a wave changes into a particle when it is "observed." Numerous experimental setups have been created that illustrate the effect of observations on subatomic particles. Einstein put his finger on a key question: how can we ask questions about objects when no measurements are made?

Such questions are complex and so counterintuitive that many physicists adopted an alternate interpretation, namely that the theory does not actually tell us about the structure of the world. That is, the theory is merely a way of computing how the world works, not describing it. Other physicists felt that describing how the world works was an important aspect of science

---

6. Woit, *Not Even Wrong*.

and objected if quietly to this interpretation. To use quantum physics to decide among different interpretations of the origin of the universe a more physical description is required. The difficulty in obtaining such a theory has been described by Adam Becker in his very readable *What Is Real: The Unfinished Quest for the Meaning of Quantum Physics*.[7] He describes how underlying and unsupported assumptions held by quantum physicists have not been examined sufficiently, leaving a plethora of conflicting theories that do not at present allow distinguishing between various interpretations. The assumptions that underly these various interpretations, for example, multi-worlds, are not based on science but rather merely on assumptions, assumptions that are seldom stated.

It does matter what scientists think about what they are doing, however, because it is their actions, the experiments they choose and don't choose to do that affect the stories that "filter out into the wider culture, changing the way we look at the world and our place in it."[8] Thus, physicists developing quantum theory tend to hold a worldview influenced by materialism and its emphasis only on things that can be directly observed. This has resulted in conflicts between quantum physics, observing only material things, and biblical descriptions of the spiritual aspects of life. As Becker emphasized, the development of quantum theory is important but unfinished, and he and others go much further into quantum physics than I can go here.[9]

As quantum theory shows, scientific exploration and the underlying assumptions can be complex. Attempting to understand the dynamics of the universe, difficult as it is, eventually challenged me to face the reality of a possible creator behind it all. But this took some time. As I began my scientific career, the first reality that I faced was that science was limited in its ability to address some of my biggest questions.

---

7. Becker, *What Is Real?*

8. Ibid., 7.

9. Wilczek gives a different look into these areas by a Nobel Prize–winning quantum physicist that is also very accessible to the nonspecialist. This is an exploration of what the author terms "spiritual cosmology," giving his answer to the question: "Does the world embody beautiful ideas?" If so, he wonders, where do those ideas come from? Wilczek, *Beautiful Question*.

7

Facing Reality: 1973–1997

A FEW DAYS AFTER I returned from Fletcher's Ice Island at the North Pole, I interviewed for a job with the Southwest Fisheries Science Center in San Diego. They needed me as soon as possible, a pattern I would grow to live with. So, I buckled down to complete my degree requirements.

My PhD committee appointed a new leader, and I revised my draft dissertation. I presented it in an hour-long public seminar. Pastor Herb and Margene came to watch, my committee came to question, and I came terrified. After the public left, my committee members continued asking questions. I hadn't expected them to be very interested, and maybe they weren't because after another half an hour they each nodded their satisfaction. Subject to fixing a few typographic errors, I had the PhD that Margene and I had sacrificed everything for. We were elated by where it might take us but also saddened by my professor's suicide. Herb had been right when he said that I got some of his pain.

It was early 1973 when we got settled in San Diego. This laboratory had recently laid off several scientists as part of a government-wide cost-cutting program, and yet here I was, the newly hired "golden boy." My training in both biology and mathematics was what the laboratory director wanted, but there was resentment among some scientists there who had just seen friends laid off. I didn't know how to respond to this mixed reception but knew I had a job to do, so I again buckled down.

My new job had been created when President Nixon in 1972 had signed a new law, the Marine Mammal Protection Act. That law demanded an answer by 1974 to the question, "Were too many dolphins being accidentally killed in the San Diego tuna purse seine fishery?" My team and

I worked out some new ways of estimating how many dolphins were in the sea using ship and airplane sighting surveys and placed observers aboard a sample of tuna vessels to count how many dolphins were being killed. In the fall of 1974, we began pulling these two sets of data together using statistical methods, some from graduate school and some new, invented on the fly.

Late one autumn Friday I settled in for a long weekend tending the bank of calculating machines that would summarize the data and then solve the algebraic equations in the population models. Those models were like those I had first seen in Thompson's book *Growth and Form* in college, and I was immensely pleased with myself and relieved. The calculations worked as planned, and by Monday morning I was in the laboratory director's office, giving him a preview of the answer: too many dolphins were being killed. We plugged the answer into our waiting draft report, made some photocopies of its 200 pages, and sent them out to colleagues for review. We had made our two-year deadline.

My First Science Report

"So, Dr. Smith, how many copies of this report did you make?" The question caught me totally flat-footed. I couldn't imagine why the man had asked this and I stumbled for an answer.

"About 20 or so, I think," I finally said.

The group all began talking at once. The man in the center chair finally pushed his chair back dramatically.

I remember the airplane landing that morning at Dulles International Airport, touching down with a bump alongside a dairy herd. I remember the long taxi ride through farmland to the city. My father had convinced me that the East Coast was all city, from one end to the other, so cows weren't what I had been expecting either. Within days of my team completing its report, I found myself bound for the agency's Washington, DC, headquarters. The laboratory director left town just after I had briefed him and was now traveling somewhere in Europe. I was flattered that he trusted me to handle this.

Now I found myself facing a semicircle of people, a few I knew of but most I didn't, answering questions. Most had not read our report, so I summarized its conclusions. All seemed to have been going well, and I remember being surprised by their deference given my lack of experience. My "golden boy" status must have come before me.

Then the man in the center stood and asked, "Couldn't we just shred them all?"

I was dumbfounded. President Nixon had just released the tape recordings that were implicating him in the Watergate conspiracy. Were they really proposing another conspiracy, I wondered? I had been trained to do science to answer questions, and here our work was in danger of being erased. "Is this really how science works?" I wondered as the memory of burning my fraudulent science fair project came to mind.

The next day I flew back to San Diego in shock, buying several drinks during the flight. I had looked into an abyss that I hadn't seen coming. In the end, the copies of our reports weren't shredded, but the covers were stamped: "This is not a document." The agency would not act on our report answering the question about the dolphins because it was explained to me, the U.S. tuna purse fishermen were politically powerful and the Nixon White House, as long as it lasted, wasn't willing to alienate them. The head of our laboratory forbad us to pursue further research on the effects of the tuna fishing on dolphins. This wasn't because of a real fire hazard, as the fire marshal had claimed about the school boiler room so long ago, but rather to protect the agency from a political fire hazard from the White House.

"This is a rare event," I explained to Margene. "The agency wouldn't do this normally."

She looked doubtful, shaking her head as I picked up another drink.

"You haven't been the same since you got back," she said. "I'm worried about you. You've been sick to your stomach for weeks now. And I'm worried about *us* because we are both drinking more."

"You mean *me*, that I'm drinking too much?" I challenged. Truthfully, I had ended up passed out on the bathroom floor at the last party we went to.

After that, I couldn't focus on where my work was going and tended to fret over details that didn't use to bother me. My PhD had taken us somewhere I had not known even existed.

Moving On

"You are lucky," the young doctor said, "you have a low tolerance for stress. Those pains will save you from getting ulcers. Can you change jobs?"

I asked the head of our laboratory about getting some relief, but he just advised antacids; that's how he dealt with his ulcers. I wasn't surprised by his response and began to see patterns. For example, I now realized that he had left me to present our report and left the country so he wouldn't have to defend it.

Finding a different place to work was the obvious way to reduce my stress, but we had moved here just two years earlier, and it seemed too soon to leave. That wouldn't look good on my resume, and besides, now that I had finished my graduate program, Margene had begun hers. She was halfway through a master's in rehabilitation counseling at San Diego State University. I couldn't ask her to give that up. But home life was not improving, and the antacids were not helping.

After two weeks of being sick, Margene gently said one evening, "You don't have to stay in this job. We can work something out about my school."

My stomach quit hurting the next day. She released me from feeling trapped by my laboratory director and by her school, and I was euphoric. Taking advice from a good friend, I did not "burn bridges" with the laboratory or the agency. I did not make my moving on into a protest against my agency's misuse of our science or my laboratory director's abandoning me; I just left.

I was invited to the University of Hawaii to interview for a teaching job, a place I had wanted to return to since my trip to Eniwetok during college. I got a job I had always dreamed of—being a professor. Margene would be able to finish her master's degree there as well, and after she finished, she got a job helping disabled people. We were riding high, accomplishing our dreams, and moving forward.

However, I was often traveling on consulting jobs to help make ends meet, traveling too often as it turned out. Margene began to develop "island fever," as the locals say. In my third year there I got into a conflict with a professor in another department over a research project, a project that he claimed rights to as a "senior professor." Teaching was fun, but I was beginning to see that the giant egos that surrounded me even in this laid-back university took a lot of that fun away. Margene and I took our home life to a counselor, and he helped us sort out what was going on. We agreed that it was time to move, again.

My PhD continued to take us on a complex career path, always following new jobs. First, we returned to San Diego, where I began trying to answer the same question that I had failed to get an acceptable answer to

Facing Reality: 1973–1997

before: are too many dolphins being killed by the tuna purse seine fishermen? The environmental community had continued to pressure our agency to address this the same question, but now we approached with a much greater understanding of the politics of science and with much-improved science. But after another three years, we got the same answer as the first time: too many dolphins were being killed. The same answer, but with many of the uncertainties now ironed out. Surely this would be acceptable to the agency.

Much had changed by then, of course. I had greater status within the laboratory and the agency because, even though I left quietly before, my resignation had been seen as a protest. I was now respected because I had been willing to walk for my principles; not everyone had. Also, President Richard Nixon's resignation had revealed the results of conspiracy, and the thinness of his justification had surfaced: "Well, when the President does it, that means that it is not illegal."[1]

But our answer was again unacceptable, even if better justified. Presidential candidate Ronald Reagan had raised funds by campaigning on the decks of San Diego's tuna purse seiners. On the television news, he promised the fishermen to get the federal scientists off their backs and their boats. Within days of President Reagan's taking office in 1981, someone from headquarters flew to San Diego and, in a hastily called full staff meeting, fired my entire research team. Instead of just having our report stamped "This is not a Document" like our first report in 1974, this time people were fired for just having contributed to our report.

There's Nothing Medicine Can Do

In the winter of 1981, Margene came down with the flu. When she went back to teaching her aerobic dancing class a couple of weeks later, she knew something had changed. Every time the dance called for a pirouette, she would stumble. Her doctor sent her to the hospital when she developed paralysis on the right side of her face. They said something about neurological testing and told her they needed to do a spinal tap.

"Oh, like for testing for multiple sclerosis," she knowingly asked having worked as a rehabilitation counselor with multiple sclerosis patients.

The test showed albumin in the fluid, suggesting multiple sclerosis (MS), but no one wanted to pronounce such a diagnosis, something that

---

1. Frost, *Frost/Nixon*.

would forever label her because there was no cure. However, based on her intermittent fatigue and loss of balance, as well as her numbness and spinal tap, the neurologist soon confirmed that diagnosis.

"There is no cure and no treatment, nothing we can do," he told Margene, deadpan, evading her eyes. "Avoid stress and come back if you have another attack. We can shorten up the duration with cortisone injections, but that's about all. Sorry."

I had never seen Margene so angry as after the doctor left. She seethed about both the diagnosis and the doctor's manner of delivering it. The bad news was one thing, but to hear that there is no cure, and especially no hope, was another. We never saw that doctor again, but his diagnosis had been correct as well as unusually quick. We would come to learn that that MS was a long-term degeneration of the nervous system. It had no known cause, highly variable symptoms, and, as the doctor had emphasized, no treatment.

And the prognosis was always bad: progressive loss of nervous control due to scarring of the insulating myelin sheath surrounding the nerves. MS was variable, but most patients gradually lost their sense of balance, muscle strength, and brain function and usually ended up in a wheelchair. We had a friend in our church who had MS, and she was in and out of a wheelchair. We had never understood what she and her husband were going through until now.

Over the next several months we began seeing a counselor, sometimes together as a family and often individually. His task was to help us adjust to the changes that MS would bring. I suppose he helped, but it's hard to point out just how. I do remember, however, being stunned by one question that he asked once when we had met alone: "So, Tim, will you leave Margene?"

I was thunderstruck and could not respond. Tears filled my eyes.

"I've counseled other young families with this diagnosis," he continued, "and often the spouse decides to leave."

His question shook me, and I barely remember my garbled response, except the point that I had never even thought of leaving. I pushed down the tears but thought about it, wondering about our marriage vows, "for better and for worse," and wondering about the limitations this would put on my life. I also wondered how Margene could believe in a good and personal God who would allow such evil. I refused to deal with these questions then but knew that I would stay.

The initial MS attack subsided, and we wondered if the doctor had been wrong. Later Margene began to experience additional episodes, gradually losing energy, clarity, and balance. She transferred her anger from that hapless diagnosing doctor to the disease itself and explored what was known and what might help. We began reading various statistical reports of different rates of occurrence of MS in different countries, in different climates, and with different diets. Right up my alley, but the statistics were always a little disappointing. Margene followed up claims that a low-fat diet helped reduce the frequency of episodes and that acupuncture helped with the fatigue. Contrary to the doctor's skepticism, some of these things seemed to help.

One day Margene explained where she had been going on Tuesday evenings for the past month or so.

"It's a charismatic Catholic women's group. Someone said people had been healed there, and I thought maybe I could be healed."

"Oh," I began, not sure of what was coming, "that sounds interesting."

"But I couldn't tell anyone at our Lutheran church. My parents warned me about Catholics."

"Last week," she continued, "I was born again." That is, she explained, she had a direct and personal experience of God, and everything she had learned as a Christian growing up "became real."

I remembered hearing Pop laugh at what he called "Holy Roller churches," but Margene's explanation was completely different from his experience and outside of my capability to imagine.

Adrift in Exile

"Excuse me; I don't understand what you said?" I asked the hardware store clerk for the third time. The store clerk's accent and attitude reminded me that we were in what seemed to be another country. So did my Western friendliness, for example, when my casual greeting to a stranger on the sidewalk in our little village was responded to with a disparaging "Do I know you?" To avoid feeling bad Margene and I frequently referenced Dorothy's line, "Toto, I have a feeling we're not in Kansas anymore."

My laboratory director in San Diego had managed to shield several members of my team from actually being fired after President Reagan took office; I never knew how. Since I had become a political liability to the laboratory, I was quietly transferred, almost exiled, to a sister laboratory in

Woods Hole, Massachusetts, where there were no problems with dolphins being killed by fishermen. We gradually settled into life in Woods Hole, our daughter started junior high, and we eventually joined an Episcopal church that hosted charismatic meetings on certain evenings like the ones Margene attended at the Catholic women's group in San Diego.

Margene began leaning into this group as her MS symptoms worsened. She hadn't gotten over her anger and offense at the doctor who said there was nothing to do for it, and she had little to do with standard medicine. Acupuncture seemed to help, at least for a while. Diet appeared to be important, and we tried many different ones. Sometimes she would get better, and yet again, she would get worse.

While Margene sought healing through her charismatic group, the Episcopal board of elders was becoming less and less tolerant of the pastor who allowed Margene's group to continue meeting in their church. Eventually, they encouraged the pastor to move on, and I was asked to chair a group to replace him. We interviewed candidates that the Episcopal synod sent our way and eventually settled on a couple to be copastors. They received approval from our elder board, albeit during a divisive meeting, and we offered them the position, which they accepted. Quickly, however, word got around the church that they, too, were born again charismatic believers.

Then I experienced the reality of something new to me: the power behind the throne. Several wealthy but not very active members flexed their financial muscles. Although there were not very many of them, they carried the type of authority that money often brings. The elder board quickly revoked its approval. A couple of us were asked to visit the couple in Boston to withdraw the church's employment offer. I chafed under this hidden authority, but I did what was asked and then resigned from the committee and the church, again in quiet protest of what I saw as untrustworthy leadership. Margene and I were adrift in a personal exile.

My research at the Woods Hole fisheries laboratory had focused on the New England cod fishery, a small but currently politically powerful group of independent fishermen dating back to before the Mayflower arrived at Plymouth Rock. Cod were being fished off eastern North America by Basques since before Martin Luther nailed his 95 Theses on the door of All Saints Church in Wittenberg, Germany in 1517.

The research group I inherited was busy with calculations about how many codfish should be allowed to be caught, and we regularly advised the fishery managers. My team's advice was to set large mesh sizes for the nets to let more fish escape. Our recommendations called for larger and larger mesh sizes as the catches declined. The theory was that a larger mesh size would make the fishermen less efficient, and hence protect more codfish. The fishermen were upset about becoming less efficient.

The analysis methods we used were different from those used in the Pacific, and I wondered why.

"I'm trying to get oriented to my new responsibilities," I said in a meeting of my team early on, "and I was wondering why you all use different analysis methods than your colleagues use in the Pacific."

Anxious to educate me as a newcomer, or to put me in my place, an uproar of voices broke out across the room of 30 or so scientists. Over the next hour, I was amazed to learn that there were many different opinions about why the West Coast scientists used poor methods. This led me to delve into the history of fishery biology, and I eventually wrote a book called *Scaling Fisheries*.[2] Although certainly not a best-selling book, it was generally well received, including by real historians—except by some of my closest colleagues.

That was when I was introduced to the idea of revisionist history, with the senior leadership of my laboratory either ignoring my book or suggesting that I had misinterpreted the history of fisheries biology. Thus, my ideas about where my group's research should go were not well received. I realized that, once again, I was contrary to my laboratory's leadership. I began to feel like I was adrift in a professional exile.

Being adrift in New England was not comfortable for West Coast people, and Margene and I began looking at jobs in other locations. Even though there would be many jobs for Margene wherever we went, there were only a few jobs in other fisheries laboratories for me. Also, I was restricted to government laboratories because my retirement wasn't transferable outside the government. Inquiries to those laboratories were not encouraging, with people mumbling about my conflicts with leadership and hiring freezes. So instead of moving on as we had too often done, we decided to stay in New England until we could retire and began to search for ways of dealing with our being both personally and professionally adrift.

---

2. Smith, *Scaling Fisheries*.

Finally Settling in New England

Personally, we had always anchored our lives in a church despite my skepticism and felt the loss from our encounter with the power behind the throne in the Episcopal church. We tried several churches, looking for a place we would feel at home. Finally, Margene heard that a young man was beginning a new church, a charismatic one. I could tolerate churches in church buildings, but this was something new to me. They were currently meeting in his home. Margene went alone to the next meeting.

"He taught about house churches," Margene explained to me the next day. "He mentioned some passages in the Book of Acts, something about a couple hosting the church. Besides, he said there were no proper church buildings for the first three centuries or so."

The latter fact caught my attention. There wouldn't have to be anyone in authority over such a church. Authorities always seemed to get in the way. I looked for the couple in Acts, mistakenly finding Peter's challenge to Ananias and Sapphira and their deaths in chapter 5. That seemed like too much authority. But Margene found another couple, Priscilla and Aquila, who hosted the church (1 Cor 16:19). They sounded safer.

We started studying Pastor Henry Blackaby's book *Experiencing God*[3] and became interested in his idea that God is working everywhere, but sometimes only shows people certain places where he is working, places where he is calling them. I was pretty leery of the idea of an active God, but Margene convinced me to at least come to the next home church meeting. We eventually settled into that church, Margene wholeheartedly, me somewhat less so.

Something very strange occurred soon after we joined that church. When Margene had been diagnosed with multiple sclerosis all those years earlier, we were told which symptoms could and likely would eventually develop. We were fortunate that the disease hadn't progressed faster, we had been told; for many it did. The balance problems had already kept her from dancing, and increasingly she was feeling fatigued. More recently her vision in one eye began to become difficult, with a screen developing that kept her from seeing well.

One Saturday morning the small group that met at our home each week came to pray specifically for her healing. This was not new and hadn't

---

3. Blackaby et al., *Experiencing God*.

seemed to help much, but that morning George, one of the men praying for her healing, became animated.

Suddenly Margene exclaimed, "I can see out of both eyes! The screen over my eye just disappeared." She described how she could now see equally well out of both eyes, the first time in a long time.

"Margene," our friend Gayle said, "I hear God saying that this is just a down payment on your healing."

I was flabbergasted, disbelieving, and afraid. I couldn't deny what Margene was saying, but I also couldn't actually see that something had really happened.

In my professional life, I began looking for funding for alternate research that would not be under my laboratory director, and I imagine he was relieved when I succeeded. It turned out that some New England fishermen did, in fact, kill dolphins in their nets, contrary to Nixon's 1972 Marine Mammal Protection Act. I went on a few more trips to headquarters, now wiser than on my first trip in 1974, and pried enough money loose to set up a modest research program. Soon I was asking the same question I had started my career with: "Are too many dolphins being killed? "

I knew too much about the likely result of providing an answer to that question, however, and instead began to broaden my research base to address additional, perhaps less volatile questions. The Basque cod fishermen provided the basis for my next question. They had also pursued whales off eastern North America beginning at least in the seventeenth century. By the eighteenth century, they and American whalers had succeeded in nearly exterminating one species of whale, the so-called "right whale." There was a lot of interest in the failure of that species and several other species to recover from near-extermination around the world. I settled into this research program, always keeping out of the way of the laboratory director. I found my professional moorings.

Over the next few years, my research program blossomed with new understanding locally of the whales, dolphins, and seals in New England. Too many dolphins were being killed, but that was addressed without causing a blowup. Seal populations were increasing, perhaps at too great a rate for the taste of local fishermen, and the historically depleted right whale population showed signs of starting to recover. That led me into the past, examining historical whaling in New England and throughout the North Atlantic and eventually into every ocean. I remember seeing the outline map of the world's oceans fill up with pink dots when we first mapped the historical whaling locations we had been assembling from the old whaling logbooks.

Things were well with Margene and me. We were in a growing church community and I was in a growing research program inside the fisheries laboratory. Life was good. But I had also experienced many events over these years that I could not explain. My success and relative comfort gave me the leisure to begin exploring wider areas of science and religion. My focus, for example, shifted from the origin of the universe to something a bit closer to home.

8

Our Cosmic Home

LIKE COSMOLOGISTS CHANGING THEIR mind about the universe, changes in ideas about the structure and age of the solar system had also contributed to my uncertainty about God when I was growing up. Just as the church on Camp Creek and the science teachers on the other side of the McKenzie River had taught conflicting ideas about the universe, they also taught conflicting things about our solar system and our earth.

The church had insisted on a literal interpretation of the words in Genesis, that the world had come into existence in six days. The science teachers had taught that the earth had always been just as we found it, for instance, that the continents were fixed and unmoving. The church's choice was disappointing because Missy, my Sunday school teacher, would not address the other possible meanings of the Hebrew words that Moses used in describing creation.

Looking back now, I've learned more about our observations of the solar system. For example, Chinese observers began naming stars and constellations around 2300 BC, and by 750 BC the Babylonians had discovered such details as a 19-year cycle in the timing of the rising of the moon. This led to the Greeks predicting eclipses by the sixth century BC.

The Bible also gives us a few clues about the form and geometry of these objects. Moses shifts quickly in Genesis from his broad strokes of beginning and creation to take a point of view from the surface of the earth. He describes God's creation of light in Genesis 1:3 and later describes in verses 14–18 the sun and the moon separating day from night and acting as signs and as markers of seasons. The sun, moon, and constellations were well known (2 Kgs 23:5), including specific constellations such as the Bear,

Orion, and the Pleiades (Job 9:9), the planets Saturn and Venus (Amos 5:26), the numerousness of the stars (Gen 13:13), the phases of the moon, and the sun moving, rising, and setting (Ps 113:3). Stars and solar system objects were described as they appeared from the earth. Thus, the sun appears to rise, but as is evident from our reliance on manmade satellites today, the earth rotates to the east, revealing the sun each day.

Although biblical authors don't spell out an image of a flat earth as the center of the universe as their Egyptian neighbors did, there are biblical passages that some have interpreted to this effect. For example, the word translated as "firmament," *raqiya,* includes the notion of flatness (Gen 1:7). Similarly, the prophet Daniel was asked by King Nebuchadnezzar to interpret a vision that included a tree that "was visible to the end of the whole earth" (Dan 4:10–11). This visibility implies, according to some, that the earth is flat.

Further, other scriptures state that the earth doesn't move. David, for example, wrote that "the world is established; it shall never be moved" (Ps 93:1). Hannah, in thanksgiving for her son Samuel, prayed: "For the pillars of the earth are the Lord's, and on them he has set the world" (1 Sam 2:8). From the point of view from the surface of the earth, these passages suggest that the sun rotates around a flat and fixed earth. The sequential development of increasingly more sophisticated models of the solar system occurred gradually over centuries, resulting in the Christian church repeatedly revising its interpretations of those scriptures. Whereas the Bible was clear that the universe had a beginning, now cosmologists began to lead the interpretation of Bible passages about the geometry of the solar system.

First, Aristotle dismantled the idea that the earth was flat with his visual observation of a curved shadow on the moon when it and the sun were on opposite sides of the earth, that is, during an eclipse. The remaining biblical picture supported Claudius Ptolemy's (90–168) geocentric model of the solar system. This model was not refuted until the early 1600s with Galileo's (1564–1642) telescope-assisted observation of the earth casting a shadow on Venus, an impossibility for Ptolemy's model, which did not allow the earth to come between the sun and Venus.

That left two possible models of the solar system: heliocentric, as argued by Copernicus (1473–1543), and geoheliocentric, as argued by Tycho Brahe (1546–1601). The latter maintained that the sun and other planets rotated around a fixed earth. It wasn't until the early 1800s, after 28 years of trying, that Fredrich Bessel (1784–1846) measured a seasonal change in the angle to a distant star, 61 Cygni. This change implied that the earth was indeed moving over the course of a year. This left only Copernicus's heliocentric model, improved by then with Johannes Kepler's elliptical planetary orbits, as the only model that fit all the observations.

Thus, contrary to geoheliocentrism, the earth is moving, and the obvious interpretations of Scripture don't tell us much about the geometry of the solar system. It is not that certain scriptures are wrong, but rather that their point of view and their purpose don't answer the question of the geometry of the solar system. Importantly, the interpretations of Scripture have had to be repeatedly adjusted as our observations of the solar system have accumulated. This is an important reminder that reinterpretations of observations are sometimes necessary and possible even though some people may continue to hold on to earlier assumptions and interpretations. The tension between new cosmological observations and new interpretations of the Bible dominated several centuries of our history and often clashed violently.[1]

Formation of the Earth

Turning from the solar system to the structure of the earth itself, Moses in the first chapter of Genesis elaborated the process of forming the earth (*'erets*), which had initially been formless and yet included water (verses 1–2). First, light was created (verses 3–5), and the light and dark periods of the earth were named *yom* and *layil*, day and night. Then some of the water was separated and drawn above (verses 6–8).

Although water on a formless earth is hard to imagine, this sequence of events is not contrary to observations by cosmologists and geologists. Water is one of the most abundant compounds in the universe, and recently cosmologists found a black hole that appears to be surrounded by enormous amounts of water.[2] Further, geologists have for years known that water today is alternately evaporating from the sea into the atmosphere and returning as rain, just as Moses and later Amos (9:6) described.

---

1. Pearcy and Thaxton, *Soul of Science*.

2. The National Aeronautics and Space Administration (NASA) maintains a website describing new developments in space exploration. The observation of water in a black hole is described at http://www.nasa.gov/topics/universe/features/universe20110722.html.

Moses continues by describing regional structures of the earth (Gen 1:9):

> And God said, Let the waters under the heavens be gathered together into one place, and let the dry land appear. And it was so. God called the dry land earth, and the waters that were gathered together he called Seas.

The image is the water forming into seas and the land becoming continents or islands. Today we can observe directly from man-made satellites that the earth is a complex, wrinkly mosaic of continents, islands, and oceans. The latter occupies 70 percent of the earth's surface, and the surface of the land and the depths of the ocean floor vary from Mt. Everest at 8.8 kilometers above sea level to the deepest part of the Mariana Trench, 11.0 kilomeers below. Further, eroding ocean shores and mountain cliffs reveal the earth is composed of distinct layers of soil and rock, often oriented horizontally but sometimes angled to the surface of the earth. How did the layers form, and why is the earth so wrinkly?

The Hebrew scriptures identify several geological processes that must have made some further changes. Strong winds and earthquakes were frequently mentioned (1 Kgs 19:11), as were volcanoes (Mic 1:4). Moving mountains were familiar although likely poetic images (Ps 114:3). Seasonal floods of the Nile were known (Amos 9:5), and that things sank under the destructive power of floods was also well known (Exod 15:5).

Geologists had these processes in mind as they attempted to understand the formation of the earth. An English farmer and geologist became fascinated with the changes in the earth itself. James Hutton (1726–1797) published *Theory of the Earth*[3] in 1778 based in part on his observations that the well-known layers of soil and rock, which had long been studied in road cuts and cliffs, were sometimes tilted. The layers suggested accumulations of different materials, and the tilting suggested that the layers were laid down or rearranged by strong, often catastrophic forces.

By 1830 another English geologist, Charles Lyell (1797–1875), published his own theory, titled *Principles of Geology*,[4] arguing that the history of the earth could better be explained by gradual forces rather than catastrophes, his theory known as uniformitarianism. This approach was adopted by many geologists, and for decades only gradual changes were

---

3. Hutton, *Theory of the Earth*.
4. Lyell, *Principles of Geology*.

considered to be plausible. Thus, the predominate layers in the earth's crust were thought to be formed by gradual settling of material.

Geologists described the accumulated layers of sediments over the next century, especially the deposits of coal, natural gas, and oil valuable as energy sources. Geology students developed ways of remembering the names of layers, beginning with Quaternary, going deeper and older through Tertiary, Cretaceous, Jurassic, and so forth back to the Cambrian, and then jumping into the Precambrian. This last jump would prove especially significant because the earth now appears to have been covered or nearly covered by ice just before.[5]

Continents Move Slowly

One of the Boiler Room Boys in grade school brought in a map showing how the continents could be fit together, like a jigsaw puzzle. We marveled by the fit between the east coast of South America and the west coast of Africa, as had been pointed out by the geologist Alfred Wegener (1880–1930) in the early twentieth century. We wondered at the forces that could have moved continents apart. We also imagined how well other coastlines fit together, but not all of the continents fit as well. Although we were fascinated, I wondered if the continents had split apart.

In junior high, I remember asking about Alfred Wegener's argument. I had looked up more information by then and knew that the idea had been around since the beginning of the seventeenth century. A Flemish mapmaker, Abraham Ortelius (1527–1598), geographer of King Philip II of Spain, had written about it when the outlines of the continents were just becoming clear in his maps. In 1912 Wegener put forward that observation again, along with supporting observations about the complementarity of the distribution of plant and animal species on the different continents, as evidence for his theory of "continental drift." My teacher was unimpressed with my question, however, and shouted at me to look at the page in our textbook that said the continents had always been where they are today. "What force could move continents?" he demanded smirkingly.

My face instantly flushed scarlet. I was mortified because I had discounted this question earlier. I wondered if my textbook was right, but I didn't dare ask that teacher again. He hadn't forbidden me to pursue Wegener but effectively stopped my questioning by posing a question of his

---

5. Macdougall, *Why Geology Matters*.

own. I felt sad for Wegener. Photographs of him had the appearance in his later years of a beaten man, but his map reconstructing Pangaea, his supercontinent, was so convincing. I was sad because I thought he had been proven wrong in something that he believed so strongly. And I was sad because I didn't have the courage to push past my teacher's question.

Geologists up to the mid-twentieth century persisted in their belief that the continents did not move, just as cosmologists had persisted in their belief that the universe had no beginning. The end of World War II and the beginning of the Cold War, dominant features of my growing-up years, provided the correction that geologists and my textbook needed.

The U.S. Navy began funding scientists to map the North Atlantic seafloor in order to learn how to detect and avoid Russian submarines. The U.S. was afraid that those submarines might deliver the nuclear bombs I had practiced hiding from under my desk during grade school.[6] One surprising observation was that as ships towed magnetometers from east to west, they detected abrupt reversals in the polarity of the mineral magnetite in the sea floor. These reversals appeared on new maps of the floor as alternating stripes, oriented roughly north to south. Even more surprising, the bands were parallel to and mirror images of each other on either side of the underwater mountain chain known as the Mid-Atlantic Ridge.

Just as the radiometric time clock in rocks is set as lava cools, so the polarity of the cooling magnetite is set to the prevailing magnetic polarity of the earth. Thus, the stripes showed how the polarity of the earth has varied over geologic time. Radiometric measurements of the age of samples collected from the sea bottom across the North Atlantic revealed another interesting pattern. Samples from near North America and near Europe were roughly 180 million years old; samples along the Mid-Atlantic Ridge were roughly one million years old, and samples from in between were intermediate in age.

The Mid-Atlantic Ridge is two parallel underwater mountain ranges stretching south across Iceland, south through the North and South Atlantic Oceans, and beyond the southern tip of Africa. Between the two ranges is a valley that can be seen today at Thingvellir National Park in Iceland. One foggy mid-summer evening my daughter Rachel and I scrambled down into the 10–15-foot-wide trench, moving slowly from the east side of the ridge and then quickly out again on the west side. It was unsettling to be inside the trench, not claustrophobic exactly, but being exposed to a

---

6. Orestes and Krige, *Science and Technology*.

geologically active trench that has been traced on the ocean bottom thousands of miles south catches one's imagination.

Rear Admiral and geologist Harry Hess (1906–1969) and marine geologist Robert Dietz (1914–1995) argued that this surprising fact of alternating polarity would be "a matter of course" if lava were emerging from the valley between the Mid-Atlantic Ridge mountain ranges. This lava pushed the continents away from each other and as it cooled became the new ocean bottom. As it cooled, it would take on the then current polarity of the earth's fluctuating magnetic field. The oldest lava was farthest from the two mountain ridges.[7] These two observations, radiometric ages of the rocks and ship-based measurements of the polarity of the magnetite, form a continuous record across the North Atlantic sea bottom. This record shows that the North American and European continents have been moving away at a steady but very slow rate of centimeters per year, now for roughly 180 million years. Further, direct measurements of the distance between the two continents using global positioning system observations have independently confirmed that they continue today to separate at a rate of centimeters per year.

These observations of the movement of continents changed the beliefs of geologists. By the time I had finished high school, geology had been turned on its head by this new explanation. I was surprised because my teacher and my textbook had been wrong. I was shocked by the tenacity with which geologists had held onto their beliefs and pleased because Wegener had been right all along.[8] These patterns and rates of movement showed that the North Atlantic Ocean had been forming slowly and continuously for at least 180 million years, a pattern that has been seen in other oceans subsequently.

Floods: Gradual and Catastrophic

The official family story was that Pop wanted to get coffee for my mother, but I always thought it was as much for his cigarettes. We had been cut off from town when, once again, the McKenzie River flooded. Floods were common enough in Oregon when I was growing up, and I remember that day watching my father walk away from us along the submerged road in his fishing waders. We waited in the car where he had parked it on high

7. Orestes, *Plate Techtonics*.
8. Ibid.

ground next to the floodwaters. An hour later he was back, carrying Mom's coffee, some candy for us, and, of course, his cigarettes. Every year the river flooded its banks, and every year the retreating water left debris.

Floods still make me nervous. The woman who often took care of my brother and me when my mother worked called me outside one misty day to see a rainbow. These were common enough in Oregon, but her explanation confused me. She said that rainbows were God's promise that he would never again try to kill everyone by a flood. I wondered why he had tried to kill everyone to begin with; why he, of anyone, had so obviously failed; what methods he might use the next time he tries.

This flood was, of course, Noah's flood, as Moses described in Genesis 6. She hadn't elaborated on the story then, but it focuses on God deciding to destroy all people: "For all flesh had corrupted their way on the earth" (Gen 6:12). Further, he hadn't actually failed in carrying out his intention but rather had allowed Noah and his immediate family a way out because they alone had been righteous. For me in my skepticism of the biblical story, this wasn't particularly reassuring. I was sure I wasn't righteous.

While the annual floods in Oregon were brief, they were frequently destructive of farmland and houses. Noah's flood, however, was different: "The waters prevailed and increased greatly on the earth, and the ark floated on the face of the waters. And the waters prevailed so mightily on the earth that all the high mountains under the whole heaven were covered" (Gen 7:18–9). Noah's flood destroyed on a much larger scale than those floods in Oregon: "Everything on the dry land in whose nostrils was the breath of life died" (7:22). The waters retreated over a matter of months, but there is no mention of the debris forming layers. Indeed, the retreat of the waters was followed by the animals and people on the ark stepping out onto dry ground. The land didn't appear to have been greatly damaged by the flood or any debris accumulated, for soon: "Noah began to be a man of the soil, and he planted a vineyard" (Gen 9:20).

There has been a complex connection between the development of geology and Noah's flood.[9] The early focus had been on explaining the number of animals the ark could hold and their migration to and from the ark. More geological concerns were addressed in the seventeenth century, such as the formation of geological layers and sufficient water for such a flood. Thomas Burnet (1635–1715) in 1684 thought that the flood waters arose from the center of the earth and retreated gradually to form geological

---

9. Young, *Biblical Flood*.

layers even on mountaintops. Robert Hooke (1605–1703) in 1688 argued that the timespan of the flood's retreat, mere months, was insufficient to lay down geological layers, and further, that the marine fossils ended up on mountain tops due to uplifts by catastrophes such as earthquakes.

Such interpretations of the relationship between geology and Scripture, especially interpretations of Noah's flood, were further developed by several popularizers of geology, one the Presbyterian Scotsman Hugh Miller (1802–1856). He emphasized that the Bible was true but not itself a reliable source of "definite physical facts, geographic, geological, or astronomical." Further, he felt that the description of a universal flood was more likely hyperbolic writing, along the lines of other biblical stories that referred to *all the world*. For example, did Joseph feed "all the world" (Gen 41:57), or did kings come to Solomon from "all the world" (1 Kgs 4:34), or were Jesus' parents caught up in a census of "all the world" (Luke 2:1)? Miller and other popularizers showed the church that biblical literalism was difficult to support with the newly developing geology.

Despite Hugh Miller's emphasis on geology over Scripture, the connection between geology and Noah's flood was revived in the early twentieth century. Seventh Day Adventist prophet Ellen G. White (1827–1915) reported visions of Noah's flood laying down the prominent geological layers. She described organic debris and corpses being subsequently buried by high winds, forming geological layers, including coal and oil reserves. Her pronouncements were treated on par with the Bible by the Adventists, and George Price (1870–1963) would soon base his teachings on them. He wrote a 1923 college textbook, *The New Geology*, arguing that all of geology could be explained by the flood.[10]

However, like many before him, the source of sufficient water for Noah's flood eluded Price. In fact, he suggested that the flood may have been due to tidal fluctuations caused by an enormous impact on the earth. There was, of course, no biblical suggestion of such an impact and no geological evidence either. Price was left to work out White's catastrophic visions based on geological observations, not Scripture, and almost all geologists of the early twentieth century were unconvinced.

Price's failure to convince geologists of the effect of catastrophic events like Noah's flood did not mean that catastrophic events were forever ruled

10. Young connects the Adventists ideas to the early nineteenth-century "scriptural geologists." He also notes that part of the focus on explaining a rapid formation of geological layers was so that Darwin's new theory of the origin of species would have nothing to explain. Ibid., 245.

out of geology. In the 1920s a stubborn young geologist named J. Harlan Bretz (1882–1981), educated and working in the Pacific Northwest of North America, also ran into a problem of a geological explanation without sufficient water. He had begun his career in eastern Washington in a very peculiar-looking area known as the Scablands.

I remember this area from camping across it during a graduate ecology course. The professor had pointed out the weird soilless landscape with enormous potholes and mesas and ripples in the sediments in the bottoms of square-sided valleys. I remember that it had a spooky feeling, nothing like the more usual Western landscape with V-shaped valleys eroded by rivers and U-shaped valleys scraped clear by ice.

Bretz had seen those square-sided valleys and in 1923 published a scientific paper suggesting that the landscape was created not by gradual erosion but by a catastrophic flood. He didn't invoke Noah's flood, but nonetheless, it would have been a flood of monumental proportions. His argument was not well received, especially by geologists trained on the East Coast, partly because catastrophic geological explanations were not in vogue.

But more substantially, just like Price's inability to explain the water for Noah's flood, Bretz couldn't explain the water for his own flood. He was initially ridiculed professionally as an upstart. Later he connected with another young geologist, Joseph T. Pardee (1871–1960), who had been studying Ice Age lakes in Montana. Pardee was more interested in ice dams on the Clark River in Montana. Those dams had formed the enormous Lake Missoula, which can still be detected in the Kalispell region of Montana. When those ice dams had eventually broken, he realized that enormous amounts of water would have flooded west into eastern Washington State.

Bretz and Pardee eventually established the importance of these catastrophic forces, but geologists were very slow to change their presumptions. Bretz was 96 when the Geological Society of America finally awarded him its highest honor. He reportedly lamented to his son on that occasion, "All my enemies are dead, so I have no one to gloat over." But his ideas were not dead, and his work turned geology on its head by allowing the effects of catastrophic flooding to be more carefully considered.

Biblically Based Estimates of the Age of Creation

In Genesis, Moses described many events of creation in a series of six passages (Gen 3:2–32). Each one begins with "And then God said" and ends with "And there was evening and there was morning, the [n]th day." These passages express the passage of time by using the word יוֹם, transliterated *yom*. How much time depends on what Moses meant by *yom* in these passages.

Historically, *yom* has been translated to mean a normal day, and the age for the earth was often computed as the six days of creation plus the number of days implied by the generations listed in the Bible's genealogical passages (e.g., Gen 5). For example, the Hebrew scholar Yose ben Halafta (second century AD) calculated that there had been at least 3,761 years between the creation of Adam and the conquests of Alexander the Great.[11] This calculation was confirmed by Archbishop Ussher (1581–1656), again by Johannes Kepler in the seventeenth century, and yet again by Isaac Newton in the eighteenth century. Everyone used similar methods and arrived at similar ages, all substantially less than 10,000 years.

At the end of the nineteenth century, such calculations lost some of their authority when William Henry Green (1825–1900) noticed that Uzziah was either Joram's son, according to the book of Matthew, or Joram's great-great-grandson, according to the book of 2 Kings. Green's observation suggested either that errors were made in the several genealogical lists or that those lists weren't intended to account for all generations.

The crux of such calculations was not the number of generations but the interpretation of *yom* to mean a normal day, making creation essentially instantaneous. There are other possible translations, and Collins systematically describes three.[12] One possibility is that Moses meant to draw an analogy between God's work and man's work. Thus *yom* is used in connection with the "evening and morning" wording to remind us of the rest periods in man's workweek. Another possibility is that Moses meant that creation occurred over a series of actual sequential time periods, but those being of undetermined length. Hugh Ross has championed this, the "day-age" view, by developing consistent interpretations of both scriptural and geological observations.[13] Another possibility is that Moses used a complex

---

11. Halafta, *Seder Olam Rabbah*.
12. Collins, *Science and Faith*, ch. 1.
13. Ross, *Navigating Genesis*.

literary structure of pairs of verses (e.g., verses 1 and 4 relating to light, etc.) embedded in a simple temporal sequence. Meredith Kline (1922–2007) championed this "literary framework" view.

Thus, Collins expressed his preference for the analogical-day view, Ross for the day-age view, and Kline for the literary framework view, and all three express their doubts of the historical normal-day interpretation. These four interpretations, of course, have very different implications about what to expect from Genesis. The normal-day interpretation places emphasis on the genealogies, with all their problems, and Henry Morris has kept this historical interpretation alive.[14] Ross attempts to harmonize present-day geological observations with biblical texts in sophisticated ways, while Collins doubts that such harmonization is needed because Genesis has different purposes than geology.[15]

These four interpretations of Moses' meaning of *yom* are entrenched, and the pros and cons are well described in many books and articles and continuous debates. There does not appear to be sufficient information internal to the Bible to resolve this question to everyone's satisfaction. Ongoing arguments of how to interpret scripture have not advanced our understanding further.

Geologically Based Estimates of the Age of the Earth

My introduction to radiation was my first pair of school shoes. We went to the Buster Brown shoe store where there was a machine that showed the bones in your feet inside your new shoes. The image was supposed to help the salesman select the right size shoes, but it didn't show the flesh of your feet, so it wasn't clear just how useful the machines were. Still, we enjoyed watching the bones in our toes move as we wiggled them. The effects of radiation on the survivors of Hiroshima were not yet widely known, and school shopping became less interesting when they were.

Radiation would eventually be used to estimate the age of the earth, but when Green raised his question about interpreting the biblical genealogies in the late nineteenth century, many other methods were being tried. These included calculations based on sea level changes, changes in the moon's, earth's, and Mercury's orbits, rates of cooling of the earth, radioactivity, and accumulation of sodium, limestone, and sediments in the oceans. These

---

14. Morris, *Science and the Bible*; http://www.answersingenesis.com.
15. Collins, *Science and Faith*, 92.

methods gave markedly different values, but most are substantially greater than the biblical genealogy-based thousands of years.[16]

For example, in the 1860s William Thompson (1824–1907), later Lord Kelvin, applied his mathematical theory of heat flow to the question. He assumed that the earth was a solid body heated by the gravitational attraction of the aggregated matter and concluded that the earth was between 20 million and 400 million years old. Thompson's calculations were undermined in 1895 by his own student, John Perry (1850–1920), who assumed, instead, that the earth was fluid inside, not solid. Using the same mathematical methods, Perry estimated that the earth was more likely two or three billion years old.

Perry's correction of his teacher was not well received. The difference in the estimates of the age of the earth wasn't resolved until 1895 when A. Henri Becquerel (1852–1908) discovered radiation coming from uranium. This was a possible source of heat that could justify Perry's assumption that the earth was fluid inside. Radiation from the element radium was soon discovered by Marie Curie (1867–1934) and Pierre Curie (1859–1906), beginning a new discipline, nuclear physics. This discipline has changed the course of history in so many ways.

In 1905 Ernest Rutherford (1871–1937) first saw the possibility of using radioactivity for estimating ages of rocks. His idea was that elements such as uranium and radium emit radiation at a constant rate, some faster and some slower. When an element emits radiation, it changes. Thus, uranium turns into lead. Knowing the rate of radiation and the levels of uranium and lead, Rutherford thought, one could estimate the time since a rock solidified. Additional details of radioactive isotopic decay were discovered in 1914, including the existence of several isotopic forms of the same elements, some radioactive and some not. By 1927 Arthur Holmes (1890–1965) measured the abundance of one isotope of uranium, denoted uranium-235, and the abundance of the lead it decayed into. From those abundances and the rate of radioactive decay of uranium-235, he calculated the age of the earth at between 1.6 and three billion years.

There were still, however, many uncertainties about the rates of radioactive decay and the accumulation of decay products that had to be investigated if geologists could be confident in such estimates. One uncertainty was the amount of decay products already present when the rocks were formed. This was solved for some radiometric decay processes by also

16. Anon., 2013.

measuring the amount of those isotopes that did not change over time. For example, the abundance of lead-207, a decay product of uranium-235, increased over time as the abundance of uranium-235 declined. But the abundance of another isotope of lead, lead-204, did not change. The ratio of the two lead isotopes could be used to adjust for the initial level of lead-207.

Another uncertainty was that the rates of decay vary with some physical conditions, such as pressure. However, it turns out that the rates vary only slightly. For example, the greatest variation is for beryllium-7, but its rate of decay varies by only 1.7 percent with increasing pressure. The variability in the rates of decay of other elements that are used for age determination is much lower. Thus, radiometric decay rates due to changes in pressure cannot have varied very much in the past.

In 1955 measuring uranium and lead isotope levels in the Canyon Diablo meteorite allowed a PhD student named Clair Patterson (1922–1995) to estimate the age of that rock at roughly 4.5 billion years. Because meteorites are debris from the solar system, that age was taken to be roughly the age of the earth, which had congealed from such debris. Patterson's estimate was substantially more precise than any previous estimate, with an error of plus or minus 70 million years. Subsequently, the estimate of the age of the earth hasn't changed although its precision has been improved to plus or minus only 20 million years.[17]

Rutherford's idea eventually worked, with estimates of the age of the earth increasing from his own of over one billion years to Holmes's 1.6 to three billion years and finally to Patterson's 4.5 billion years. These estimates are completely at odds with an age of thousands of years based on an interpretation of Moses word *yom* as 24 hours. This suggests that other choices of how to interpret *yom* would be more consistent with the results of geological observation and analyses.

However, the details of Patterson's calculations may not be convincing to some, and there are many books written about their reliability. There is a simpler, if much less precise, observation that may help. First, the rates of radioactive decay for each isotope can be expressed in terms of the time it would take half of a sample to decay. Many isotopes have very short half-lives, several with half-lives less than 70 million years. None of those isotopes have been found in the earth. Thus, the age of the earth must be much greater than 70 million years to have those isotopes disappear completely.

---

17. For a description of methods of determining the age of the earth. Dalrymple, *Ancient Earth*.

Indeed, after 20 half-lives, or 1.4 billion years, a small but detectable residue of a 70-million-year isotope would still exist (actually 0.0001 percent). Such ages are less than Patterson's 4.5 billion years, as would be expected because of the simplicity of this method, but nonetheless, these ages are still far greater than the biblically derived ages of thousands of years based on genealogies.

Interpreting the Earth

My babysitter's assurance that God promised never to flood the world again did not reassure me about God's nature. Further, the insistence of geologists that continents did not move convinced me that geology was unlikely to be interesting science. Looking back now, I am disappointed I didn't pursue that science; so much happened when the geologists let go of their assumption that the earth did not change.

We have many more tangible observations about the origin and development of the solar system and of the earth than we have about the universe. We observe the repeating patterns in the movements of planets, and we make direct measurements of global processes like ice ages, flooding, and continental drift. Further, we can analyze the chemicals of the solar system using meteorites that fall to the earth and from the rocks and minerals of the earth itself.

These observations, taken together, give us a greater ability to test our understanding. The interpretations of geological, astronomical, and theological observations have only partly converged, and that has happened slowly, sometimes over centuries. Competing interpretations have continued, such as whether the solar system and the earth are thousands or billions of years old. The difference between these interpretations comes from inconsistent understandings of how the biblical texts relate to the astronomical and geological observations.

While there is still ambiguity in interpreting all these observations about the solar system and earth, it is far less than the obscurity of the universe; that ambiguity is something I needed to be prepared to live with. Dealing with ambiguity, however, whether in science, religion, or life generally, would not prove easy. This came home for me as my boundaries of belief weakened when someone outside my understanding began breaking into my life.

9

Coming Home: 1998–2005

I COULDN'T PUT MY finger on it, but by early 1998 things were distinctly changing in our lives in New England. One thing I noticed was seeing many strange quotes from famous people about God. I usually loved these, pretending that I understood something about the person from their one-liners. My page-a-day calendar included C. S. Lewis's metaphoric description of himself as a fox being pursued: "The fox had been dislodged from the Hegelian Wood and was now running in the open . . . bedraggled and weary, hounds barely a field behind."[1]

This reminded me of the title of a poem I later learned was written by a failed seminarian and opium addict, Francis Thompson (1859–1907). His autobiographical "The Hound of Heaven" describes him fleeing God "down the labyrinthine ways of my own mind, and in the midst of tears I hid from him."[2] Thompson's full poem revealed that he also felt like God was hunting him down. I found a recording of Richard Burton reading it, a reading that brought me to tears. What was happening to me? I realized that I desperately wanted science, not God, to be in control.

Then another unexplainable event. During a training session about how to pray for healing that Margene had dragged me to, a woman we only vaguely knew prayed for the healing of Margene's fatigue. It had gradually gotten worse and was increasingly dominating our lives. The woman began crying as she prayed. Margene said she felt more joy than she had before the prayer. But there was nothing to suggest any healing.

---

1. Lewis, *Surprised by Joy*, 226.
2. Thompson, "Hound of Heaven."

Coming Home: 1998–2005

However, when we sat down to dinner the next evening, Margene asked if I noticed anything different. I had: she was unusually alert for this time of day. She said she forgot to take the second dose of a medication she depended on twice daily to give her energy. Instead of being wiped out, as I would have expected, she was still alert. The difference between my expectation for her and her reality was unbelievable. She felt that it was due to the prayer the evening before, but I wondered.

I knew that the medication, an amphetamine, stimulated the production of dopamine, a neurotransmitter essential for neurons to pass electrical impulses to the brain. I knew that the amphetamine had been overcoming the effects of the gaps in Margene's myelin sheath, the sclera visible in neurological scans that multiple sclerosis gets its name from. But to have a sudden change so that the amphetamine was no longer needed, *that* was beyond anything that science could explain, except by the catch-all phrase "went into remission." I did not understand, and perhaps I did not believe even though this time, unlike when she said her vision had been healed, the reality was obvious even to a doubter. I was getting increasingly confused.

Another quote from C. S. Lewis came up on my page-a-day calendar, about his realization, just before he accepted that there was a God, that he wasn't alone in his battle of belief, comparing it to playing Solitaire (the card game Patience for him). He wrote, "I realized that there is another Party in the affair—that I'm playing poker, not Patience, as I once supposed." I remembered again playing two-handed Pinochle with Pop years ago, just to be with him, not because I liked card games. I knew that even when he dealt me a strong hand of cards, eventually he would win. There was nothing I could do but sit there as he revealed his hand, step by step, and respond from my hand, card by card.

That reminded me of how much I preferred our childhood games from Monopoly to Pinochle, even though my father still almost always won. The difference between Monopoly and Pinochle is stark. In Monopoly everything is laid out on the table, there are no secrets, and things happened predictably, if not always in my favor. In Pinochle all the cards are hidden and predicting what will happen is nearly impossible—except I almost always lost.

In the middle of all these uncertainties, I began to meet with our pastor. Initially, I tried to explain my fears to him, what was happening, hoping he could explain all of this. But other issues arose. He began speaking to me about our church incorporating so we could buy a building that would allow for more Sunday school rooms. The argument for buying began to look sketchy to me as members started to leave and the financial offerings began to decrease. He had asked our parent church for their approval to incorporate, but his supervising pastor disapproved, saying we shouldn't try to do this yet. He told me about this during one of our meetings and then asked very pointedly, "Tim, would you and Margene be willing to come with me if I pulled away from our parent church and became an independent church?"

I hadn't seen this coming and resented being put on the spot. I tried to sidestep by discussing the nature of authority within churches, something that I truly did not understand. That didn't help. While I had long been uncomfortable with authority, doing my best to avoid being under anyone, the pastor turned out to have even more problems with authority. He continued to prod me with his question about following him. I could see that my answer was important to him, so when I finally told him that we were unwilling to go along with withdrawing from the authority of our parent church, he was visibly upset. I didn't know yet just how upset he was.

A few weeks later he asked several of the church's leaders to meet one evening, as usual at our home. Since we had last spoken, he had gone ahead with the idea of incorporating the church. He found a lawyer who volunteered to create the necessary paperwork, and that evening the lawyer presented those documents.

I was stunned. Where he had felt uncomfortable with the authority of our parent church, I now felt uncomfortable with his.

"I can't sign this until we've had more opportunity to discuss it as a group," I finally said, trying to defuse what looked like an escalating conflict. "There are budget problems developing, and people have begun to leave."

"Are you with me or against me, Tim?" he asked, turning toward me and pointing in the same way he had when he asked me this question before.

"I am with you," I said. "You are my pastor. But I can't agree with this. Members are leaving, and our reserve fund is shrinking rapidly. We don't have our parent church's approval."

"Tim, either you are with me, or you are not," he countered, raising his voice to a near shout as he attempted to assert authority that I realized I could not accept.

I retreated into our kitchen, leaning my head to the cupboards, tears of pain flowing. I didn't know if I had done the right thing or if I was just responding to my own fear of authority.

Word and Spirit

Knowing this conflict had been difficult for me, Margene arranged for us to get away. I had long avoided going to the Christian conferences that she favored, but this time her appeal caught me at a weak moment.

"We wouldn't have to attend all the sessions," she went on with a suggestive wink.

"Perhaps we might have some time alone," I thought, "away from the church." I agreed to her suggestion. We would attend the Word and Spirit Conference in Connecticut at the end of the month.

It was a three-day conference, longer than most that Margene went to, and I was resigned to attending at least some of the many sessions.

We went to the first session of the conference that evening, a worship service followed by a speaker. I liked the Word part of the conference title, but I wasn't so sure about the Spirit part. Worship began loudly that evening, and then, suddenly, I saw a clear picture in my mind: a large hand was holding a contract, with the thumb over the bottom where the fine print usually goes.

Then I heard a calm, but firm voice ask, "Yes or no?"

I knew this was God, and I knew that it would be dumb to answer, "No."

"Yes," I heard myself say aloud.

The moment struck me as somehow familiar, reminding me of an evening long ago when I was a young boy: "Everyone, please close your eyes, so no one will be embarrassed," the pastor of Camp Creek Community Church had said nearly half a century before. "Those who would like to give their lives to Jesus, please raise your hand."

I remembered that night like it was yesterday, all the details. The hateful green mohair sweater that I wore that evening, my only sweater. After we raised our hands, peeking to see who else was doing this, we repeated some words after the pastor. I had wondered for a while what was supposed to happen. Later, as I realized that nothing had changed, I gradually forgot about it.

Now in Connecticut, I wondered if something had changed back in Camp Creek when I raised my hand. Had God taken me up on my earlier response, so that now all these years later I was able to hear his question, "Yes or no?" I also wondered if my recent conflict with our pastor had shaken me, challenged my confidence enough to shake me.

In any event, I did hear that voice, and I did say "Yes," nearly leaping up to answer my second altar call and dragging a surprised Margene down to the front. Where I had come to this conference hurting with despair about things from my past and our current church, now I felt joy and freedom. Something had lifted off of me, enough so that I lost my usual reticence. We even happened to find the pastor whose firing had driven us from our Episcopal church years before, and he and his wife prayed for us.

We did not tell our current pastor what happened that weekend. I was still too tender to share with him and there hadn't been time before we left for a long vacation to the family blueberry farm in Oregon.

Margene and I hadn't lived there since before we were married, and our daughter had never lived there. Nonetheless, that was home for all of us, and our daughter, who had just graduated with her master's degree in archeology, wanted to be married there. I had arranged to use three months of accumulated annual leave that summer, and it would be a time of preparing for a wedding, reading, picking blueberries, and fishing in the McKenzie River over the hill behind the farm, where I had once killed a rabbit for its skull. It had been a generation ago, and now I was returning to where my two roads, science and religion, had diverged. I began watching them converge.

I had been surprised by God and now wondered if mine was a weird story. Sometimes I wondered if it had just been some psychological perturbation. An English biologist friend had published a book describing his encounter with a goddess, a decidedly non-Christian and unconvincing goddess. Others wrote of their own encounters with God, but nothing I had read sounded like what I had experienced. There was no consistent answer, so, being academically oriented, I began to read in earnest.

I started with C. S. Lewis, a convert to Christianity from atheism, and continued with many other authors, trying to answer the many questions that were arising. The obvious question was what I had said yes to. I hadn't read the contract that I saw in my mind's eye; hadn't been able to and hadn't felt I needed to. But now what really worried me was what the thumb had covered, the fine print that so often includes specifics that are not so joyful. I had experienced something that was outside my previous experience, and I did not even know what questions to ask.

For example, Lewis described his own conversion more concretely, using the analogy of a cat and a mouse. In a collection of his letters he wrote, "Terrible things are happening to me. The 'Spirit' or 'Real I' is showing an alarming tendency to become much more personal and is taking the offensive and behaving just like God."³

I understood now that this meant that there is something that is sometimes present, like what Pastor Herb said might have been Jesus just off to my left when our daughter was born. I wasn't sure about the difference between the Spirit and Jesus and God, but I thought of what I had seen disappear that day when I killed the rabbit for its skull. Hardly reassuring to an atheist, but then I guessed I wasn't one anymore if I ever really had been. One change was that now I thought I understood the fifth line of the old hymn "O Holy Night," usually reserved for Christmas: "Till he appeared, and the soul felt its worth." This was the "something more" I had experienced, what Lewis described as the "Real I."

In his poem "The Lockless Door," Robert Frost (1874–1963) described someone escaping through a window adjacent to the unlocked front door after inviting in whatever or whoever had been knocking:

> It went many years,
> But at last came a knock,
> And I thought of the door
> With no lock to lock.
>
> I blew out the light,
> I tip-toed the floor,
> And raised both hands
> In prayer to the door.
>
> But the knock came again.
> My window was wide;
> I climbed on the sill
> And descended outside.
>
> Back over the sill
> I bade a "Come in"
> To whatever the knock
> At the door may have been.

---

3. Lewis, *Collected Letters*, letter to Owen Barfield, February 1930.

> So at a knock
> I emptied my cage
> To hide in the world
> And alter with age.[4]

I realized that it was I who thought he had tricked God, always hiding from him in plain sight. Staying close to God, walking alongside—but not in—the river of the Spirit, as Margene and Gayle called it, had been a safe way to protect me from something unfathomable, from being caught in a card game instead of my preferred game of Monopoly. I realized that I had been stepping over that windowsill for a lifetime, trying to keep control of my world. But that evening in Connecticut I finally chose not to do that. I gave up my control by submitting to the authority I had so long evaded.

For decades I had ignored experiences that science couldn't explain, unwilling to face the possibility that God existed. I had pushed the idea of him away when he was standing beside me by our daughter's incubator. I had ignored my roommate's pain even though I had seen it. I had sidestepped my professor's suicide, running to the North Pole. I had discounted Margene's reports of healing from MS, which threatened my understanding that science alone was enough. I had ignored my deceit about my high school science fair project. I realized that I had not wanted God to be in charge; science was much safer. Many of these experiences I was now able to give to God in exchange for forgiveness that I found through his grace.

While I had long believed that science would one day explain everything, including who I was inside, I now realized that not everything could be explained merely by science, which brought relief. As one Nobel laureate and agnostic physicist put it, "Science isn't everything, thank goodness."[5] Although I wanted God to be unnecessary, he was revealing that there was so much more to the universe, to the world, and to life than I had any idea of. I had wanted what the leader of an atheist and freethinkers group had said when he wrote a short statement that summarized the teachings of his "religion." He answered: "As we are averse to believing things with no proof, and as there is no proof for the existence of supernatural beings, we aver that no gods exist."

The archaic *aver* means to "assert to be the case," and that leader said he knew of no proof of God's existence. I realized that there is likewise

---

4. Frost, "Lockless Door."
5. Wilczek, *Beautiful Question*.

no proof for God's nonexistence. Thus one can just as reasonably aver that God exists as that he doesn't. These become alternative beliefs; one or the other must be true. But mere logic itself doesn't settle this. My encounter with God changed my belief, but *my* experience isn't the proof an atheist requires. Even the reports by so many others of their experiences of God aren't sufficient. Those reports, however, encouraged me that I hadn't merely fallen out of sanity or rationality.

I could now understand the testimonies of people about having been encountered by God, testimonies that I had so long and so easily discounted. The many ways people reported being encountered by God began to haunt me. C. S. Lewis's card game and Francis Thompson's flight were but two of those reports. Many others described God encountering them suddenly or gradually, gently or violently, directly or indirectly. Behind each such statement was a personal experience, one that changed each person.

I also began to see the many encounters reported in Scripture, stories I had dutifully read or had been preached to me for years. In Acts, for example, Saul of Tarsus was struck down by God's question, "Why are you persecuting me?" This encounter included Saul being blinded by a light and later healed (Acts 9). Later two women and a jailer were encountered by God through the same Saul, now renamed Paul. One heard Paul's gentle words, another heard his annoyed words speaking harshly to a fortune-telling spirit, and the third heard his loud words interrupting an attempted suicide (Acts 16).

I began to look for patterns in these reports, a natural history of people's collisions with God. But the patterns were neither in the settings nor the words. The pattern was in the outcome: people changed after God's encounter. I knew I had changed, but when I discussed my experience of God with Pop, he described his rejection of God after he had attended a church once as a young man.

"The same people who attended that church would be out whoring around the next week, pretending to be holier than thou," he said. "People never change."

And that is another part of the pattern. People do not change merely by someone's report of God's encounter or by someone's persuasive words. People change because of their own encounters. Proof for the atheists and freethinkers would have to be much different from such testimonies, and they certainly wouldn't fall for a tricky "logical" proof of God's existence like the ones I remember from high school, something like: "God must exist because otherwise, we couldn't even ask the question."

Such linguistic tricks aside, the often-mentioned proofs of God's existence are interesting. These arguments apply three logical premises to interpret what we can observe about the universe, our world, and ourselves. The Kalam premise, mentioned earlier, is that everything that begins to exist must have a cause outside of itself. The teleological premise is that the extreme suitability of the universe for life could not be explained by random chance but only by design. The moral premise is that objective moral values are only possible if there is a source.

Three arguments then proceed by applying each premise to the observations we have made. The observation for the Kalam argument is that the universe had a beginning, the Big Bang, and hence must have had a beginner. The observation for the teleological argument is that neither physical necessity nor chance has been shown adequate to explain the observed suitability of the universe for life, and therefore something must have designed the universe—and us. The observation for the moral argument is that people everywhere are consistent in that they consider some actions to be *right* and some actions to be *wrong*. That consistency implies there must be an independent source of that knowledge.[6]

These three arguments had not convinced me of God's existence because they depend on their underlying premises being true. I had been comfortable in just denying the premises, which are not proven, and nullifying the corresponding argument. After God encountered me, I was convinced that he exists, and I was able to reexamine these premises. Now they carry greater logical weight, and now they tell me some things about this encountering God.

Further, those three arguments for his existence tell us something more. The Kalam argument would not only confirm his existence but also that he is not restricted to our universe. This had not been clear to me growing up. The God of my childhood Sunday school was also in this world, just like me. But the Sunday school paintings of God on a very physical-looking throne in the sky had bothered me: how did heaven stay up there, what with gravity and all?

The teleological argument would confirm that he is creative in carefully designing the universe. The general description of this is that the universe is surprisingly "fine-tuned" as a habitat for life. This is seen in the precision

---

6. Craig, *Does God Exist?*; Sire, *Apologetics of Reason*.

required of our mathematical models of the universe. These models imply that very small differences in physical constants would prevent life from existing. Given the many precise requirements, it appears extremely unlikely that such a universe could have formed by chance.

Finally, the moral argument would confirm that God is ethical. My experience as a child of WWII showed me that people could act in evil ways, including the Nazi terrorism but also the responses to that evil, bombing German and Japanese cities and killing many civilians. I had understood then that measuring good and evil was thought of as a matter of numbers of people and the values given to each. I now saw that God values all people equally and individually, so those simple evaluations are not sufficient for deciding what is right and what is wrong.

Although I my immediate thought was that I had encountered Yahweh, the God of Moses, I later began to wonder. Could I have collided with some other god? How would I know? Based on Huston Smith's (1919–2016) *The World's Religions*, what I had experienced did not seem like the gods described by the Eastern religions: Hinduism, Buddhism, Confucianism, and Taoism. Further, those descriptions of god didn't match the implications of the Kalam or teleological or moral arguments. Similarly, Aldous Huxley said we could only approach his "Mind at Large" through systematic meditation, something I had not done.

But if I were right and I had encountered Moses' Yahweh, what should I make of Jesus? I recalled when I found Jesus mentioned by Josephus in his *Antiquities of the Jews* that had convinced me that Jesus had existed and especially that Christianity originated in the first century. Reading further, now I found in book 18, chapter 5 of Josephus's *Antiquities* his description of Herod's fear of "John, that was called the Baptist." Josephus wrote:

> Herod, who feared lest the great influence John had over the people might put it into his power and inclination to raise a rebellion, (for they seemed ready to do anything he should advise,) thought it best, by putting him to death, to prevent any mischief he might cause.[7]

The story of Jesus' baptism by John the Baptist occurs in the four separate first-century writings that are primary documents in the Christian Bible, the books by Matthew, Mark, Luke, and John. They all share the image of the Holy Spirit descending "like a dove" when John reluctantly

---

7. Josephus, *Antiquities*, book 18, ch. 5, para. 2.

baptized Jesus in the Jordan River, and three of those documents share the well-known words: "This is my beloved Son, with whom I am well pleased" (Matt 3:17). Those words feel like what I felt when God had asked, "Yes or no?" He didn't say that he was well pleased with me, but I felt then his acceptance of me despite my reluctance.

The fourth book, written by Jesus' disciple John, reports John the Baptist's testimony (John 1:32–34):

> I saw the Spirit descend from heaven like a dove, and it remained on him. I myself did not know him, but he who sent me to baptize with water said to me, "He on whom you see the Spirit descend and remain, this is he who baptizes with the Holy Spirit." And I have seen and have borne witness that this is the Son of God.

This tied up loose ends for me because, first, I also knew from Josephus that John the Baptist had lived and had been killed by Herod. Second, the image of a dove from Missy Lane's Sunday school was replaced by the Greek word *pneuma*, Spirit, that was like a dove, but had remained on Jesus rather than flown away. Third, John continued, this all signified that Jesus would baptize with the *hagios pneuma* or Holy Spirit. Fourth, John states that Jesus is *huios ho theos*, the son of God. I knew that God had spoken to me, and now I knew that this involved Jesus as the son of God and the baptism by the Holy Spirit. I also knew that I had a lot to do to comprehend all of this.

Could I Have Answered No?

If matter is all that there is, as my early science claimed, then we should be able, at least in concept, to describe everything that happens in terms of that matter. But my sense that I chose to say yes raises the question: could I could have chosen to say no? I believe that God, who is not matter, chose to ask me a question and that I had the freedom to answer either way. My first thought was, "I know this is God. It would be dumb to say *no*." Dumb, yes, but was it even possible? Philosophers and theologians have discussed this question for centuries, and lately, even some scientists have added to the discussion, albeit with little to contribute. And the answer is strangely elusive.

"See," my brother Eric had taunted back in the Boiler Room days, "people are just things and cannot choose." Eric was pointing at a passage in a new science fiction book that described humans as:

> A self-balancing, 28-jointed adapter-base biped; an electro-mechanical reduction-plant, integral with segregated stowages of special energy extracts in storage batteries, for subsequent actuation of thousands of hydraulic and pneumatic pumps, with motors attached; 62,000 miles of capillaries; millions of warning signal, railroad and conveyor systems . . .[8]

I was intrigued by the mechanical imagery but thought there must be more. I felt there was something inside me that was looking out.

But if I didn't, in fact, have to choose my answer, if it was already embedded in my brain by past events and experiences, then why would my answer have changed anything? My mind revolts against my answer being not really my choice, and I have to agree with neurophysiologist Sam Harris (1967–) about the significance of this feeling of having free will:

> The question of free will touches nearly everything we care about. . . . [M]ost of what is distinctly human about our lives seems to depend upon our viewing one another as autonomous persons, capable of free choice.[9]

"Autonomous persons" raises a possibility that my brother Eric and I had explored, that we were such persons while other people were robots. That meant that their responses to us were predetermined. But if all our responses to the universe are based on what has gone before and none are based on our free will, then how could we generate anything new, for example, knowledge? It would seem more likely that our predetermined "choices" would lead us toward some shadow of reality determined by previous events.

The Unexpected Fine Print

I also began to see how I could have participated in Christian churches for years, first as an atheist, later as a theist, and finally as a Christian. In C. S. Lewis's *Surprised by Joy*,[10] he described a similar journey with two

8. Fuller, *Nine Chains to the Moon*.
9. Harris, *Free Will*.
10. Lewis, *Surprised by Joy*.

abrupt changes in his life that sounded similar to changes in my life. The first occurred on the top floor of a bus traveling up Headington Hill, when he grudgingly and suddenly gave up his atheism. The second occurred in the sidecar of his brother's motorcycle on the way to the zoo at Whipsnade, when he moved from theism to Christianity, becoming "the most dejected and reluctant convert" to Christianity. About this he wrote, "Really, a young Atheist cannot guard his faith too carefully." My Headington Hill occurred gradually after high school, and my Whipsnade Zoo occurred when God asked me, "Yes or no?"

Eventually, I came to understand that the contract I had said yes to consisted of one word: freedom. There was nothing else on the contract, not even where the fine print usually goes. I was reminded that Paul had written to the church at Galatia (Gal 5:1): "For freedom, Christ has set us free." Then I understood why God had pursued me: he has me where he wants me, and regardless of the contract guaranteeing my freedom, I want to—and will—do what he is calling me to. As Paul wrote to the church at Ephesus (2:5–10):

> Now God has us where he wants us, with all the time in this world and the next to shower grace and kindness upon us in Christ Jesus. ... God does both the making and saving. He creates each of us by Christ Jesus to join him in the work he does, the good work he has gotten ready for us to do, work we had better be doing.[11]

Saying yes to God left me wondering what the freedom he promised means and consequently, what life was really supposed to be. This opened a new world of questions, both about spiritual life and biological life. Even the biology that I knew, when touched by God, was beginning to take on new dimensions.

---

11. Peterson, *Message*.

10

Science, Religion, and the Life In-Between

I AM STILL HAUNTED now several decades after high school by what happened when I cut that wounded rabbit's throat. Watching the rabbit's eyes, I saw something disappear when it died; animation ceased, yes, but something more. I recognized the presence and then the absence of life without knowing what life really was. Life didn't come along with the skull that I took back to the laboratory. What disappeared remained a mystery, something beyond the matter, the headless body that I hastily buried that day, ashamed. Since that day, biologists have learned much more about what life is and what it is not.

The first animated movement that Moses described in Genesis was "hovering": "And the Spirit of God was hovering over the face of the waters." The second was "sprouting": "And God said, 'Let the earth sprout vegetation.'" The nature of vegetation is not described, and the connection between the earth and sprouting is left dangling. Many more animated movements followed: "And God said, 'Let the waters swarm with swarms of living creatures, and let birds fly above the earth across the expanse of the heavens.'" Between the hovering, sprouting, swarming, and flying, Moses described the formation of the land and the appearance of the sun and the moon, which defined the repeating patterns of day and night, of seasons, of years. The animated movements also repeated, but in ways that were qualitatively different from the repetition of the seasons.

Repeating animated movements and other functions like metabolism and reproduction are characteristic of most forms of life. Operational definitions of life are commonly given involving such characteristics, but these tend to be long and involved and, in the end, not comprehensive. There is always something missing operationally, and there is always something missing at a deeper level. This latter was often described as a "vital force," analogous to Newton's mechanical force, that animates living organisms.

But what is it that animates life? Indeed, what is life? Moses did not attempt to answer this question, and some suggest it is a futile question, using an analogy of the difficulty of defining water before chemists learned its chemical composition. This analogy suggests that when we eventually learn the composition of life, a definition will be obvious.[1] Herbert Spencer wrote of life as the "connexion between the changes undergone and the preservation of the things that undergo them." While Spencer gives no mechanism for the connection, others have used a more abstract analogy of life as connection, such as the Hare brothers, who said, "Life is the hyphen between matter and spirit."[2]

Origin of Life

We don't know what life is, but we do know much about it. In 2010 Kenichiro Sugitani and colleagues traced life back in time, identifying microfossils in Australian rocks back 3.8 billion years.[3] Sugitani's microfossils have amazed most biologists, as reflected by J. William Schopf's (1941–) surprise: "No one had foreseen that the beginning of life occurred so astonishingly early."[4] Life's sudden appearance suggests that a suitable environment formed soon after the formation of the earth 4.5 billion years ago.

All life has very similar chemistry, composed of the elements of the periodic table that the Boiler Room Boys painstakingly copied. The basic chemical is carbon, although silicon has similar properties that some think could form a basis for a different form of life. All life is built of left-handed amino acids, although right-handed versions exist that some think could form a basis for a different kind of life. What such right-handed or silicon-based life forms might be like is not known because they have not been found.

All life exists in nonliving structures called cells, separated by walls. Life functions through many proteins that are synthesized inside of cells. The one organic molecule that is necessary for all life that we know of is

1. Clelland and Chyba, "Defining Life."
2. Hare and Hare, *Guesses at Truth*.
3. Sugitani et al., "Biogenicity."
4. Schopf, *Cradle of Life*.

Science, Religion, and the Life In-Between

deoxyribonucleic acid, or DNA for short. Discovered in 1869, its biological role was gradually elucidated over the next century. This molecule, now known to be in the shape of a double helix, determines the biochemical processes that make the proteins necessary to maintain life. The two strands of this molecule unwind in certain cells when an organism reproduces, carrying the information necessary for the next generation. Surprisingly, DNA does not make the proteins itself. Rather, it includes sequences of chemical building blocks that tell cells how to make the proteins required for life to continue.

These sequences are codes for making proteins, a process that is analogous to writing. The idea I want to communicate is in my mind, and I construct words, letter by letter as I type, to convey it. The words are not the ideas themselves, but rather patterns of ink that we have agreed carry certain meanings. Based on that agreement, which is different in different languages, our ideas are conveyed to other minds. Similarly, the proteins necessary for life are not the codes in the DNA, but rather the cell reads the codes and thereby "knows" what proteins to make next. Sequences of chemical codes in DNA are the genes that we often allude to when speaking of heredity. The detailed study of the structure of DNA has improved our understanding of how life works and has allowed biologists to manipulate organisms. For example, DNA segments from one organism are now routinely inserted into the DNA of another, modifying that organism in clever ways. While such study is answering many questions about life, it has not addressed how life began. Further, it raised another question: where did the system of genetic coding itself come from?

There are three common explanations for the sudden appearance of life. First, Moses described God creating life, starting with plants and then animals, and the intricate functioning of DNA suggests that he designed life. Second, some biologists suggest that early conditions were such that nonliving chemicals spontaneously assembled into some primitive form of life. For example, in 1871 Charles Darwin[5] wrote to his friend J. D. Hooker with a big *if*:

> But if (and oh what a big if) we could conceive in some warm little pond with all sort of ammonia and phosphoric salts—light,

---

5. See Peretó et al., ""Charles Darwin," for a fuller description of Darwin's beliefs on the origin of life.

heat, electricity present, that a protein compound was chemically formed, ready to undergo still more complex changes ...⁶

Third, in 1880 the lawyer turned geologist Otto Hahn (1828–1904) claimed that fossil organisms were present in meteorites he had examined.⁷ Hahn's subsequently disproven claim formed the basis for the idea of panspermia, that somehow life originated elsewhere and was transported to earth soon after it formed.

These three possible explanations could require huge leaps of faith. First, Moses' explanation would require a worldview that allows for a creative God. Second, an "infinitely improbable"⁸ random chemical event might have occurred given enough (perhaps even an infinite amount of) time, but its occurrence very soon after conditions became suitable flies in the face of what we mean by random chance and infinite time. Third, the interstellar transport of life also seems improbable, unless life is abundant in the universe. But in any event, this would answer only the question of life's appearance on earth, not its origin. For that, we have only the first two possible explanations.

Aristotle thought that although life reproduced through eggs and seeds, it also continued to originate anew spontaneously. For example, biologists thought that what we now know as microbes emerged spontaneously in decomposing vegetation or meat or even mud. Similarly, plagues of rats were thought to emerge spontaneously, as the writer Alexander Ross stated in the mid-seventeenth century: "To question this is to question reason, sense, and experience. If he doubts of this, let him go to Egypt, and there he will find the fields swarming with mice, begot of the mud of [the Nile], to the great calamity of the inhabitants." However, spontaneous generation of maggots in rotting meat was disproved in 1668 by Francesco Redi (1626–1697) when he experimentally excluded flies from meat and got no maggots.

Darwin's "warm little pond" and Hahn's panspermia were picked up again in the mid-twentieth century by J. B. S. Haldane (1982–1954) and underlay modern-day searches for the origin of life.⁹ Experiments with

6. Darwin, *Darwin Correspondence Project*. Letter to Hooker, February 1, 1871.

7. See several articles in *History of Geology*, at http://blogs.scientificamerican.com/history-of-geology.

8. Eccles used the term "infinitely improbable" to describe the likelihood of one's specific genome (10^{-15000}), but the phrase works here. Eccles, *Evolution of the Brain*, 248.

9. For example, see Harvard University's "Origin of Life" project. https://origins.

Darwin's pond were conducted by Stanley Miller (1930–2007), which did generate some amino acids and, perhaps, suggested a first step in creating life. This caveat was dropped by the author of my high school biology text, who claimed Miller had actually created life. Although the conditions of their experiments have long since been shown not to mimic the early earth of more than 3.8 billion years ago, their results are still regularly cited as evidence that we are close to creating life chemically. Miller repeated his experiment over the next several years, and recent analyses of some of his unanalyzed data revealed even more amino acids than he had thought, but those results are thought by many to be irrelevant concerning the creation of life.

Elaborating this general approach, in 2003 Loren Haarsma and Terry Gray speculated that life began by chance in a three-step process:

> First, in the right environment (hypotheses include underwater thermal vents, shallow surface ponds, sandy beaches, volcanic craters, clay deposits, and weathered feldspar), simple organic molecules concentrated and *self-assembled* into strings of nucleic and amino acids (RNA and proteins). Second, when enough of these molecules were concentrated together, they formed an *interacting auto-catalytic system* that jointly catalyzed their mutual reproduction. Third, these RNA-and-protein catalytic systems evolved, with RNA and eventually DNA taking on the role of *information storage*, which we see in all living cells today.[10]

Despite the wealth of hypotheses about possible environments where life might have originated, none of Haarsma and Gray's three steps has yet been observed. No new "self-assembled" strings of anything have been seen assembling; no new "autocatalytic systems" have been seen interacting, and no new molecular "information storage" systems have been seen storing. The likelihood of such complex physical processes resulting in any of those steps occurring by random chance would obviously be very low, perhaps discouraging research in this area.[11]

Given the obvious difficulty in demonstrating the origin of life on earth, Haldane's other approach to the origin of life, panspermia, has

---

harvard.edu/.

10. Haarsma and Gray, "Complexity," 297 (emphasis added).

11. There have been attempts to prompt scientists to focus on the origin of life using offers of prize money. In 1999 the nonprofit Origin-of-Life Foundation reportedly offered a $1,000,000 prize and in 2011 Harry Lonsdale offered a $50,000 prize. Apparently, no one has successfully claimed either prize. Bhattacharjee, "Scientist-Politician-Atheist."

become a favorite rationale for the search for life on other planets. If life were abundant in the universe, then there may have been sufficient time available on other planets, if not on this one, for random chemical processes to have resulted in the formation of life. Researchers have been scanning the skies with radio telescopes for more than 50 years, trying to detect signals from intelligent life. Recently, the focus has been on possibly habitable planets in other solar systems. Today as I write, a group called Search for Extraterrestrial Intelligence proposed to send messages to other solar systems advertising our presence to whatever forms of life might be out there. Similarly, some scientists are examining sedimentary rocks on the earth and from Mars for evidence of life, and others are scanning primitive planets for evidence of self-replicating but nonliving molecules, another possible forebearer of life.

Despite this research, we have not witnessed the origin of life on earth nor the presence of life elsewhere in the universe. This suggests that something is missing from these approaches. For example, neurologist John Eccles's comment about the human genome would also apply to the origin of life itself: "There is a Divine Providence operating over and above the materialist happenings of biological evolution that eventually resulted in the creation of the human genotype."[12] Such was the first of the animated movement Moses described in Genesis, the Spirit of God hovering.

While we can manipulate and re-form life in many ways today and will undoubtedly be able to do even more so in the future, we have a long way to go to understand it, much less create it. This reality has been caricatured in many scientist-making-man jokes, all ending with some variation of God saying, "Get your own dirt." This snappy but too oft-repeated joke could mean that biologists cannot create life, but it could also mean that even if we could create life, we couldn't create a planet or a universe that is essential for life. Either way, it pokes fun at the hubris some people carry so easily.

That hubris is expensive when science creating life is put forward as if it were a foregone conclusion. That we don't know how life began is the truth; that that gap will soon be filled by biologists is wishful; that that gap is explained by "God did it" is less than illuminating. The tension is how to learn how life began without making claims that are merely a result of assumptions.

12. Eccles, *Evolution of the Brain*, 251.

Science, Religion, and the Life In-Between

Variety of Life

Moses followed his description of the first life on the third *yom*, "Let the earth sprout vegetation," by distinguishing "plants yielding seed according to their own kinds, and trees bearing fruit in which is their seed, each according to its kind." He then goes on in the fifth *yom* to describe the origin of marine and flying animal life: "*sherets nephesh chay*," translated "swarms of living creatures." *Nephesh* refers here to "breath," and *chay* means "life." In the sixth *yom*, Moses identifies terrestrial animals, using "*nephesh chay*" again, and identifies livestock and creeping things and beasts.

All the forms of life Moses described as being created according to their מִין, transliterated *miyn,* meaning "kind." *Miyn* is not restricted to mean a species, which we usually define as a group of organisms capable of reproducing together. Thus, *miyn* is not used when God blesses his created life to reproduce: "Be fruitful and multiply and fill the waters in the seas, and let birds multiply on the earth" (Gen 1:22). Rather, *miyn* is a more general term, one that is not simply defined.

Moses described the origin of animal and plant life in marine and terrestrial habitats with great generality but little specificity. There are only 81 named animals and only 36 named plants mentioned in the entire Bible. But life occurs in far greater variety. Although we don't have a count of all species, there are estimated to be roughly 11 million in total. Of those, we have identified only 1.4 million.[13]

One obvious division is grouping life based on their form, for example into plants and animals, as Moses described. A different approach is grouping similar species into genera, and grouping these together, for example, into families. Between families of genera and the kingdoms of animals and plants, many classification systems have been suggested. Life is not spread uniformly over creation like a bed of moss. Rather, organisms are distinct from one another and occupy distinct habitats, but in that variety, they share many similarities in their physical makeup.

Differences among species range from small morphological differences between similar species to enormous differences between, for example, the smallest and the largest life forms. The size of the DNA molecule, about the diameter of a human hair (three nanometers, 3×10^{-9}m) sets a logical lower size limit on the size of organisms. The smallest living things that have actually been observed are viruses, ranging from tens to hundreds of

---

13. Mora et al., "How Many Species."

nanometers (about 1/1,000 of the thickness of a sheet of paper). Viruses are invisible to optical microscopes because the wavelength of visible light is in the range of hundreds of nanometers. Life occurs in larger but still single-celled organisms, such as amoebas at 300,000 nanometers (three sheets of paper). Multicelled organisms are much larger, with 25 percent of all land animals being ants, ranging from one to 50 millimeters, or millions of nanometers. The largest animal is the blue whale, at 30 meters in length (or 30 billion nanometers). Flowering plants range from the tiny duckweed at thousands of nanometers to the largest tree, the redwood, at 100 meters in height (or 100 billion nanometers).

Life occupies a wide range of habitats also, extending down into the pressure of the ocean depths and up to all but the tallest mountains. Life flourishes in temperatures from below freezing to above boiling, the latter remarkably in the recently discovered hydrothermal vents at the bottom of the sea. Life has proven able to adapt to all manner of habitats and to colonize newly formed habitats. For example, I remember in high school in 1963 when the island of Surtsey emerged out of the Mid-Atlantic Ridge just south of Iceland. I agreed then with what turned out to be poorly formed speculation that it would take a long time for plants and animals to colonize this sterile lava island. But, in fact, flying insects were there within one year, plants within two years, and ten plant species had established themselves over the next 20 years.

There were many species of organisms available to invade Surtsey, and that diversity is a direct consequence of life's essential property of renewal and diversification. While chemical elements have specific, unchanging forms, life renews itself in ever-changing forms.

Patterns of Diversity

In addition to the patterns in diversity in the fossil record, biologists in the nineteenth century were finding other examples of diversity. One way was through exploratory expeditions. Three nineteenth-century expeditions were particularly important. Captain FitzRoy's (1805–1865) voyage of the HMS Beagle from 1831 to 1836 provided a platform for the naturalist Charles Darwin. Moritz Wagner's (1813–1887) collecting expedition in North Africa from 1836 to 1839 explored two rivers running into the Mediterranean Sea from the Atlas Mountains. Alfred Wallace's (1823–1913) collecting expeditions between 1854 and 1862 in the Malay Archipelago spanned the Lombok Strait in Indonesia.

Each of these expeditions revealed that closely related species were separated over relatively short distances by geographic barriers. Darwin described finches varying in bill length and shape on the different Galapagos Islands. Wagner found very similar species of a genus of a flightless desert beetle (*Pimelia*) on each side of his rivers. Wallace found very different species of plants and animals on the Bali and the Lombok sides of the 35-kilometer wide Lombok Strait.

Another diversity was variability in the bodies of the same species of plants and animals as revealed by husbandry. Selective breeding has been practiced for a long time, with beagles being developed in the fourteenth century, for example. Such breeds persist, however, as a beagle won the New York Westminster Kennel Club Dog Show in 2008.

But selective breeding has a much longer history. For example, Moses in Genesis tells the story of his grandfather Jacob tricking his deceptive father-in-law Laban. The story in Gen 29 hinges on the diversity of the herds of sheep and goats that Jacob cared for on Laban's behalf to earn Jacob's daughters in marriage. As Jacob prepares to leave with his brides after many years of caring for Laban's flocks, he requested payment for his labor. Asked by Laban to name his price, Jacob asks for all the spotted or striped or black animals. He then separates these from the white ones by having his sons pasture them several days apart.

Then Jacob puts sticks with the bark peeled in strips in front of the stronger, not the weaker, of Laban's remaining white animals. Jacob's belief was that what an animal sees when breeding would affect the coloration of its offspring. By this device, Jacob expected the stronger white animals would produce spotted or striped offspring, which Laban had agreed would be Jacob's. Apparently, Jacob's strategy worked, and he eventually left with Laban's daughters and his augmented flocks, much to Laban's disgust.

The spatial, temporal, and somatic diversity of life had been well described by the mid-nineteenth century, just as the diversity of the solar system and the galaxies beyond was increasingly well described in the twentieth century. The fossil record was revealing the historical diversity and, along with collecting expeditions, also revealing spatial diversity. Cross-breeding of domestic animals and plants was revealing the range of differences within species. The patterns were awe-inspiring and begged to be explained.

Biological Layers

In AD 60 St. Paul survived being bitten on the hand by a poisonous snake that emerged from a campfire on the Mediterranean island of Malta (Acts 28:2–7). The islanders thought Paul was a god because he had survived. Related to this story, "tongue stones" that were readily found on Malta were widely credited with supernatural powers, especially against poison. This resulted in a medieval trade in what were then mysterious stones.

In the seventeenth century tongue stones were determined not to be supernatural. The Danish priest and geologist Nicolas Steno (1638–1686) observed that these stones were very similar to teeth from a recently caught shark and argued that they had been turned to stone by some unknown chemical exchanges.[14] Steno thought that the oldest of the geological layers that had revealed so much about the formation of the earth must occur deep in the earth. He also thought that those layers could be used to establish a chronology of organisms, the fossil record.

Mary Anning collected and sold fossils as curiosities to tourists visiting the Dorset coast of England in the early nineteenth century, well before Steno's interpretation of fossils was generally accepted. When she was only 12 years old, she extracted a skeleton of an *Ichthyosaurus*, a marine reptile, from the sea cliffs, and later skeletons of a *Plesiosaurus*, a *Pterodactylus*, and a *Squaloraja*. She found these and many more fossils in specific layers of the eroding shoreline and supported her family by selling them to gentlemen collectors and scientists.

In the 1820s another woman also collected fossils, one Mary Ann Mantell, the wife of the country doctor and naturalist Gideon Mantell. She reportedly found a fossil tooth that was later used to establish the reality of dinosaurs, this first one being named *Iguanodon*. By the end of the nineteenth century 30 relatively complete skeletons of that dinosaur had been discovered in a 1,000-foot-deep coal mine under the town of Bernissart, Belgium. This left no doubt as to the authenticity of Mary Ann Mantell's discovery.

The developing understanding of the fossil record through the collections by Mary Anning and Mary Ann Mantell and others began to challenge the interpretation of Moses' description of the creation of the many forms of life. *Iguanodon* and other dinosaurs were not mentioned by Moses, so perhaps Genesis was not complete. Further, fossils of the same species from

---

14. Cohen, *Noah's Flood*.

different layers were not exactly the same, so maybe the forms weren't fixed as was often assumed. Although Moses had written that plants and animals would reproduce according to their own kind, using the word *miyn*, he had not said that the kinds were fixed at creation.

A modern interpretation of the fossil record has identified many other forms of life that have no obvious connection with anything Moses described. Further, the occurrence of fossil species in distinct layers in the earth suggested that they were not all created at once, but rather gradually and sequentially and with variation from layer to layer.

Today geologists and biologists date the fossil record from the first signs of life, the microfossils, beginning roughly 3,800 million years ago. For some 3,300 million years life was "curiously uneventful,"[15] with life occurring as single cells. In 1868 the Scottish geologist Alexander Murray (1810–1884), working in Newfoundland, discovered the earliest obviously multicellular organism, the circular ecardia, *Aspidella terranovica*. This was followed around 500 million years ago by a proliferation of fossils laid down during the so-called Cambrian Explosion. Subsequent extinctions and proliferation of life forms culminated very roughly 250 million years ago when the fossils that Mary Anning and Mary Ann Mantell were finding were laid down, the Triassic and Jurassic Periods. Around this time the earliest mammals appeared in the fossil record, and the Atlantic Ocean was being formed as Europe and North America separated along the Mid-Atlantic Ridge, as discussed above. Another mass extinction followed around 65 million years ago, wiping out the dinosaurs, apparently caused by an asteroid striking the earth near Chicxulub on the Yucatan Peninsula, and giving rise to the so-called age of mammals and birds.[16]

Explaining Diversity

Moses' explanation of the creation of individual forms of life (*miyn*) had long been the basis for Christian explanations of diversity. The varieties of organisms that might be classified as the same *miyn* were not described, although Jacob's husbandry of his father-in-law's flocks suggests an awareness that changes can occur within sheep and within goats and certainly,

---

15. See the *B.C.* cartoon for July 18, 2015 at http://www.JohnHartStudios.com for evidence of Precambrian life being uneventful.

16. Macdougall, *Why Geology Matters*.

that sheep and goats were distinct organisms. By the nineteenth century, biologists were beginning to extend such ideas to changes between species.

The most famous theory explaining diversity was Charles Darwin's 1859 *Origin of Species*,[17] in which he explained the temporal and spatial patterns of diversity using the ideas in a 1798 book by Thomas Malthus titled, *An Essay on the Principle of Population*.[18] Malthus's observation was that plants and animals, including humans, had a tendency to produce more offspring than can survive and that there were natural processes that caused them to die, for example, a shortage of food. This resulted in a struggle for existence. Although Malthus was especially interested in the social implications of this struggle for people, Darwin and others focused on the implications for plants and animals.

The implications were enormous. Small differences between parents and offspring could be preserved from generation to generation, as shown in plant and animal husbandry. For example, dog breeders have established many "breeds" by artificial selection of traits, ranging from Great Danes to Chihuahuas. If differences that favor one individual in a group over another individual were preserved in nature, then the frequency of the favored individuals in the group would increase. Under some circumstances, the group might gradually accumulate enough differences to be seen by biologists as a different species. Similarly, such new species might, by the same mechanism, change so much as to no longer appear to be related to other similar species and to be seen by biologists as a new genus.

Darwin first described this possibility in an unpublished 1844 essay. There, he argued, the existence of higher organisms can be explained by selection among "infinitesimal varieties" via the operation of "death, famine and struggle for existence," and that this process "could produce infinitely numerous organic beings, each characterized by the most exquisite workmanship and widely extended adaptations." He congratulated himself on his "view of life":

> There is a [simple] grandeur in this view of life with its several powers of growth, reproduction and of sensation, having been originally breathed into matter under a few forms, perhaps into only one, and that whilst this planet has gone cycling onwards according to the fixed laws of gravity and whilst land and water have gone on replacing each other—that from so simple an origin,

---

17. Darwin, *Origin of Species*.
18. Malthus, *Essay*.

through the selection of infinitesimal varieties, endless forms most beautiful and most wonderful have been evolved.

But why did he not publish this grand view of life when he first wrote it? That same year (1844) Robert Chambers (1802–1871) described another idea about species arising from other species in his anonymously published *Vestiges of the Natural History of Creation*.[19] Chambers, a Scottish journalist, wrote the book in Edinburgh, reportedly while recovering from a mental illness. The book had the effect of popularizing the idea of progressive species change, especially among social radicals and the "lower classes" because it suggested a progressive social change. However, Chambers provided no mechanism for such changes and additionally made many strange statements that were clearly untrue. His book was not well received by biologists, and the idea of progressive species change became a sensitive topic.

One early reader of Chambers's *Vestiges* was Alfred Wallace, who had yet to begin his explorations of the Malay Peninsula mentioned above. Wallace thought that the book presented "an ingenious hypothesis" that would serve as "an incitement to the collection of facts and an object to which they can be applied when collected."[20]

While collecting in the jungles of one of the Malay Islands in February of 1858, Wallace became sick with "intermittent fever." He later wrote that in his feverishness he recalled Malthus's descriptions of the checks to population growth, and "there suddenly flashed upon me the *idea* of the survival of the fittest—that the individuals removed by these checks must be on the whole inferior to those that survived."

Wallace described working out the specifics of this theory over two feverish hours. He drafted a manuscript over the next two days that he titled, "On the Tendency of Varieties to Depart Indefinitely from the Original Type." He sent it to Darwin in the next mail, forcing Darwin to finally publish his ideas in order to claim that he had thought of natural selection first.

Why Darwin had been reluctant to publish his ideas was perhaps because he expected a strong negative public response to some of them. Chambers, when asked why he had published *Vestiges* anonymously, voiced that same concern when he responded, "I have 11 reasons," indicating his

---

19. Chambers, *Vestiges*.

20. Quammen describes these events and especially the intrigue that followed over who would be credited with the idea of natural selection. Quammen, *Song of the Dodo*.

children. Darwin, as well, had children and important social standing. Wallace, however, had neither and "nothing to lose."

Wallace's paper and a rapidly assembled paper by Darwin using some extracts from his 1844 essay were presented together in 1858 at a scientific meeting, engineered by Darwin's colleagues to establish Darwin's claim to the idea. Darwin went on in 1859 to publish his groundbreaking *On the Origin of Species by Means of Natural Selection, or the Preservation of Favored Races in the Struggle for Life*.[21]

Natural Selection?

Darwin's and Wallace's theory of natural selection claimed that the unexpected observation of the great diversity of life would be "a matter of course" if natural selection operated randomly on sufficiently variable organisms. The best-adapted forms would gradually emerge via survival of the fittest.

However, biologists had already made other observations that were not explained by this theory. One was that species are distinct in the fossil record with few intermediate forms. Darwin admitted this but suggested that it was a transitory problem due to the incompleteness of the fossil record. Second, biologists knew that many new life forms appeared suddenly in the fossil record rather than gradually; Darwin claimed that further research would also solve this problem.

Third, biologists knew that some complex organs needed to be complete in order to function and so thought that such organs couldn't develop gradually. Darwin agreed that "It seems, I freely confess, absurd in the highest degree" that the eye could have been formed gradually. Nonetheless, he then went on to suggest how it might have happened as if merely supposing a plausible mechanism was sufficient proof.

Fourth, very similar organs occur in obviously different species, such as the eye of the human and the octopus. Darwin's slow, random process of natural selection might have produced such eyes once, however unlikely. But, it was even more unlikely that natural selection could have done this *twice*, and such eyes are frequent among different types of animals. Darwin explained this as happenstance, despite its unlikeliness, drawing an analogy to two men happening on the same invention simultaneously.

Darwin's answers to these four concerns were not convincing to most biologists of the latter half of the nineteenth century, and without

---

21. Darwin, *Origin of Species*.

the advocacy of Thomas Huxley, who early applied this argument to man,[22] Darwin's (and Wallace's) theory would have lost support much earlier than it did. But by the end of the nineteenth century, this theory was all but abandoned.

Heredity

A bigger problem for Darwin than just the uncertainty inherent in the fossil record was the then lack of a convincing theory of heredity that would explain how variability arose, was transmitted from parents to offspring, and was preserved from generation to generation. In addition to Jacob's biblical cross-breeding story, other mechanisms had been proposed. Jean Baptiste Lamarck (1744–1829) published in 1801 his *Theory of Inheritance of Acquired Characteristics*.[23] He argued that changes to a parent's body during its life, that is, acquired characteristics, could be transmitted to offspring. Herbert Spencer echoed Lamarck's theory in his 1864 *Principles of Biology*,[24] also arguing that the variability of offspring was due to what their parents were exposed to during their lives.

Struggling to understand how his natural selection might actually work, by 1868 Darwin had picked up on Spencer and published two ideas to support his theory of natural selection. One idea was pangenesis, that minute particles of inheritance, called gemmules, would be modified over the course of an organism's life. Those changed particles would accumulate in reproductive organs. The second idea was blending inheritance, that the variety inherent in the accumulated gemmules would be averaged in the next generation.

Just as Einstein a century later would be wrong when he asserted that the universe had no beginning, Darwin was wrong about both pangenesis and blending inheritance. But whereas Einstein changed his belief when Hubble demonstrated that galaxies were moving away from the earth, Darwin held his belief for decades in the face of strong counterevidence, which included tests of pangenesis by his cousin Francis Galton (1822–1911) and tests of blending inheritance by the Austrian monk Gregor Mendel (1822–1884).

---

22. T. Huxley, *Evidence as to Man's Place*.
23. Lamarck, *Theory of Inheritance*.
24. Spencer, *Principles of Biology*.

The Boiler Room Boys

In 1870 Galton published the results of transfusing the blood of a white rabbit to a black rabbit. If Darwin's gemmules were present in the blood, then he expected the offspring of the transfused black rabbit to be different from those of untransfused black rabbits—they weren't. Darwin discounted Galton's results, however, suggesting that his gemmules might exist in other bodily fluids than blood.

Mendel was an Austrian monk who, contrary to Darwin, was convinced that species were fixed. Nonetheless, he was interested in variability from generation to generation. In 1866 he published the results of cross-breeding garden pea plants, where he kept track of traits, such as the color and shape of pods and seeds, in successive generations. Contrary to the simplicity of blending inheritance, Mendel showed that such traits were passed in pairs, with one dominant trait of a pair usually being apparent and the other recessive trait usually being masked. No evidence of averaging variations was seen. Darwin didn't respond to Mendel's results even though his paper was widely circulated and cited some 14 times by others before the end of the century. Because of the persistent belief in blending inheritance by Darwin and other biologists, however, Mendel's results were ignored.

Another blow was that Wallace parted ways with Darwin's theory of heredity when August Weismann (1834–1914) showed in 1883 that germ cells (eggs and sperm) and somatic (body) cells were independent. This confirmed Galton's rabbit transfusion experiment, showing that the variability of traits developed during the parents' lives could not be passed to their offspring. Also, many biologists were unconvinced that blending inheritance in averaging variable traits could maintain variability from generation to generation. Averaging was known to move things to the middle, removing variability.

Darwin frequently explained away uncertainties using phrases like, "There is no difficulty in supposing that."[25] His decreasing number of followers also tended to use this approach, sometimes sketching very speculative family trees. One commentator later criticized this development, noting, "Such and such a character was or might be adaptive was regarded by many writers as sufficient proof that it must owe its origin to Natural Selection." Such speculations were used, he went on, "to plant wildernesses of family trees over the beauty-spots of biology."[26] Such arguments became

---

25. Collins, *Science and Faith*, 260.
26. J. Huxley, *Evolution*, 22–23.

like Rudyard Kipling's (1865–1936) *Just So Stories*,[27] a fanciful children's book first published in 1902 about how some animals acquired particular characteristics, like "How the Whale Got His Throat."

Two Evolutionary Syntheses

The eclipse of Darwinism that I had read about in college in Julian Huxley's book *Evolution: The Modern Synthesis*[28] was not the end of his discussion of natural selection and evolution. In this first synthesis, Huxley argued that the weaknesses in Darwin's formulation had been overcome by the mid-twentieth century and the theory of natural selection was alive and well. He distinguished three epochs of explaining the patterns of diversity of life: Linnaean, Darwinian, and Mendelian. Linnaean explanations in the eighteenth century were that species were fixed and created in place. Darwinian explanations in the nineteenth century were that all species had descended from one or a few common ancestors via natural selection. And in the twentieth-century Mendelian explanations involved genetics, albeit also with natural selection.

Mendel's cross-breeding of garden peas had been "rediscovered" and repeated for other species at the beginning of the twentieth century, demonstrating the particulate nature of inheritance. Weismann's germ cells were followed by demonstrations of cell divisions and chromosomes. Large and small mutations were documented, and recombinations of chromosomes were observed. The origin and transmission of genetic variation needed for natural selection were demonstrated.

But just as Mendel's particulate heredity was much more complicated than Darwin's pangenesis and blending inheritance, so the variety of mechanisms resulting in Mendelian species formation was more complex. Several ways in which Darwin's natural selection might actually work have been suggested. Thus, Darwin's quest for the "origin of species" might better have been titled in the plural, "origins."

But even as Darwin and Wallace's insight about natural selection seemed to Huxley to have been basically correct, he was careful to distinguish what had been observed from the mechanism producing it. Except for a few fossil lineages such as horses, for most groups all we have observed are "species, subspecies and genotypic variants as they exist in the present,

27. Kipling, *Just So Stories*.
28. Huxley, *Evolution*.

for those are the only groups with concrete biological existence. These obviously represent the *results* of evolution, but often tell us little about its past course."

This distinction today is made by distinguishing micro- and macroevolution. The former, microevolution, is natural selection of randomly occurring genetic variation and has been demonstrated especially with short-lived microorganisms and, indeed, even with Darwin's original Galapagos finches. The latter, macroevolution, is what is left over after microevolution explains as much of the diversity of life as it can. One limit on microevolution is the infinitely unlikely random and repeated emergence of obviously similar organs, like the human and octopus eye, through small gradual steps. The demonstration of, rather than the speculation about, macroevolution was left by Huxley as an unresolved problem.

How macroevolution occurs has been pursued by biologists developing and expanding many new disciplines. An obvious example is the direct analysis of DNA, allowing the tracing of the underlying codes of life over the supposed gradual course of natural selection. Another example is the possibility of nongradual change due to nongenetic inheritance. Biologists have brought to bear so many new ideas and techniques that in 2010 Massimo Pigliucci (1964–) and Gerd Muller (1953–) edited a book of independently authored chapters that they titled, rather self-consciously mimicking Huxley, *Evolution: The Extended Synthesis*.[29]

Unlike Huxley's, this book has not been uniformly well received. The editors describe at length in a first chapter, and in many separately published editorials, how the newer biological tools and disciplines both relate to and especially extend the *Modern Synthesis*. Detractors argue that there is nothing new in this synthesis and supporters argue that the earlier synthesis is now replaced. This disagreement reflects the possibility that declarations of an *Extended Synthesis* are premature, but the attempt illustrates the much wider range of ideas about how the variety of life may have developed.

Darwin and Wallace's initial theory was that the diversity of species would be "a matter of course" if natural selection operated gradually on heritable genetic diversity. Julian Huxley's *Modern Synthesis* was similar but with a better understanding of the generation and heritability of genetic diversity, but it still kept the underlying assumption of a gradual process operating on random variation. It is here that biologists were stuck for more than a century. Just as cosmology expanded rapidly after Hubble's

---

29. Pigliucci and Muller, *Evolution*.

observation of an expanding universe was adopted, so biology may be able to proceed in explaining the diversity of life as the developments in the *Extended Synthesis* is taken on board.

Interpreting Life

We know many things about how life functions. We have observed that the biochemistry of life is similar for all forms, implying that life diversified from a single beginning. We have observed single-celled fossils in rocks over three billion years old, very early in the history of the earth. We have observed a proliferation of types of multicellular organisms, beginning with fossils in rocks roughly 500 million years old. We have observed that the timely synthesis of many unique proteins within the cells of living organisms is regulated by a biochemical coding scheme embedded in the DNA molecule.

However, while we can recognize life and when it disappears, biologists do not understand how life began nor, indeed, just what it is. The simple definition by Herbert Spencer that I encountered in college, that something was alive if it made internal changes in response to external changes, has proven inadequate, as have any of the many other attempts. For example, the authors of an article titled "The Definition of Life: A Brief History of an Elusive Scientific Endeavor," conclude that a definition is unnecessary: "if the emergence of life is seen as the stepwise (but not necessarily slow) evolutionary transition between the non-living and the living, then it may be meaningless to draw a strict line between them."[30]

Scripture describes the Spirit of God (*ruwach*) hovering and God (*elohiym*) creating life. God spoke in the plural: "Let us make man in our image" (Gen 1:26). It is tempting to interpret this as defining human life as a connection between matter and spirit, as a hyphen. There are still gaps in biology and in theology in our understanding of life. We simply don't know what life is. Progress in our understanding of life is impeded when we agree to plug the gaps with faith in an uncertain interpretation of Scripture or with faith in some future yet ill-defined biological study. Isaac Newton spoke about this tension when he said, "Tis much better to do a little with certainty and leave the rest for others that come after you than to explain all things by conjecture without making sure of any thing."[31] Although we

30. Tirad et al., "Definition of Life."
31. Wilczek, *Beautiful Question*, 81.

don't know what life is, would we live it better if we could define it? I suspect not, because we tend to live our experiences rather than our intellect.

As I began to prepare for retirement, I thought a lot about living out the freedom that God promised in the contract that I had said yes to. I was leaving a career in science and was picking up the alternate path, that of religion, that I had long ago turned from. Reconciling the ambiguity in understanding these two paths would draw me much closer to God than I ever expected.

11

Living the In-Between: 2006–2010

TENDRILS OF FOG ENTANGLING the arches of the Bourne Bridge made me sad that early November morning when we left Cape Cod westbound. We had crossed that bridge in the opposite direction more than 20 years earlier, never dreaming that we would stay so long. Although we had adapted over the years, we realized that New England wasn't fully our home. Now, newly retired, we set off in our small camping van, named after Aslan, C. S. Lewis's Jesus character in his *Chronicles of Narnia*. We expected this to be a long road trip. Most of our extended family lived near the family farm in Oregon, and there was a lot of road between here and there. November wasn't the ideal time for a camping trip, but we had felt an urgency to leave soon after we retired, though we couldn't quite put our finger on why.

Our only scheduled stop was on the next day. Margene had read a book by a pastor from northern California who was speaking at a town in Connecticut, not far down Interstate 95 from Cape Cod. Margene wanted to hear him, but I was indifferent. The conference speaker was unsettling. His sermon cadence was one pithy statement after another, punctuated with pauses as we attempted to take in his meaning. His message rang true, and yet I was suspicious. Margene and I agreed that we wouldn't fit into the large church that he claimed, but why were we even discussing that?

Aslan began taking us on what would be a 3,000-mile trip, first further south and then, as the weather improved, north again. We visited several churches as we continued west, Aslan always taking us toward the warmth, and we found ourselves evaluating each one. We were looking for something, and although we didn't know what, we were feeling pushed by the search. We continued west through New Mexico, where we saw real mountains again and spent a week in a campground just north of Tucson. When we came to the Pacific Ocean, we turned north and followed Interstate 5 into California's Central Valley. One evening we found a campground along the Sacramento River in the small town of Redding, California. Margene remembered that the church of the preacher we had heard speak in Connecticut was in this town.

"We could just visit a few minutes," she suggested, recalling my reticence about the preacher.

"Well, we could, but we need to be getting on," I stalled, knowing that she would win.

We found his church easily, conspicuously high on a hill over Interstate 5. As we approached, I felt that this was a mistake. The road up the hill was lined presumptuously with flags from many countries. However, the building itself wasn't large, and the church office was quite modest. Still, I was glad when we left an hour later, heading north over the Siskiyou Mountains on now snow-covered Interstate 5. Aslan downshifted as we began the steep climb out of the Sacramento River Valley.

Our arrival at the family farm in the Camp Creek valley just before Christmas revealed that everyone was in transition. My younger sister was considering buying that little farm. My mother herself was considering selling it and buying a house near my other sister. Our daughter and her family had just moved from Colorado to a nearby community. We were asked what we were planning to do in retirement. While most of the family knew their plans, we didn't have anything nearly as concrete.

Over the next several months we explored the Pacific Northwest, camping up and down Interstate 5. We began to wonder what we had gotten ourselves into, indeed what we were trying to do. Gradually we realized that we were expecting some direction, but felt no draw to any place. During a church conference, a retired Coast Guard officer started to pray for me but suddenly stopped.

"I think you've been expecting God to speak to you about where you should go next but are disappointed that you've heard nothing. You are afraid you've missed what he said."

Immediately I was in tears, realizing that he was right.

"But our God is not the type of Father who lets his children miss what he is saying. When he speaks, you will hear him."

We continued to the Oregon coast, dropping south towards what we hoped would be warmer and drier places. Just south of the Oregon border it was still raining. To get out of the rain, we turned east over a complex series of mountains, and in March we arrived in the Sacramento River Valley, again at Redding. We returned to the same campsite we had used the first time we were here, several months back, along the Sacramento River. It was warm, so we relaxed. There were many geese and ducks in the lake adjacent to our campsite, along with a jet-powered boat with large red letters on its hull: "Redding Fire Department." The crew on board was tormenting the geese by testing its water cannon when we arrived.

Margene wanted to go to the prayer chapel at the church we had been to before, and though I was uneasy about it, a quick visit seemed safe enough. Later we went back to our campsite and made sandwiches for lunch.

As we were eating, Margene said, "God spoke to me today when I was in the prayer chapel."

"Oh," I managed, immediately on edge. "What did he say?"

"That I could fly here. I . . . I was," she hesitated, "I was afraid to tell you."

Her words felt cold in me. I remembered that she had shared wanting to fly at a church conference several years back although she didn't know exactly what that meant, something about freedom. Now her words seemed to carry weight, if not clarity.

I heard the Redding fireboat rev up. Looking up, I saw its bow lift, accelerating immediately to full speed. But it was headed directly toward the shore separating our backwater from the river itself, looking like it would crash into the round river rocks just 100 yards east of us.

I pointed toward the impending disaster and burst into tears, but then the boat just vanished. Suddenly I heard the words, "Just go for it," in my mind. I realized that that was God's invitation to go with Margene's invitation. I suddenly knew we were to be here. I wasn't happy, but I was relieved to have some clarity. We walked over to see what had happened and found a narrow brush-lined stream just inches deep flowing toward the main channel of the Sacramento River. The fireboat had been able to negotiate this tiny stream because it had no propeller, just the twin jet thrusters.

By July we arrived back in Redding with our rented moving truck carrying much of my mother's furnishings that she did not want to move from her farmhouse. It was 116 °F and 10 percent humidity. "What have we done?" we asked each other. "Mom said it was hot here, but *this* hot?"

The War Between Science and Religion

"I'm so sorry," a young woman whose name badge said 'Ruth,' quietly interrupting my lecture at the church's ministry school, "but in my church in Alabama we never discussed why the pastor taught that the earth was created 6,000 years ago and my high school science teacher said it was billions of years old. When I asked at church, it felt like a forbidden subject. I hoped that maybe here . . ." she trailed off.

We had settled into our new home in Redding and joined the church of the pastor we had heard in Connecticut when we started our camping adventure. I had gradually overcome my initial reticence and found that this move ended up providing a place for me to work out the changes in my beliefs that had occurred after I answered yes to God's question. One issue that concerned me and some in our new church is the relationship between science and religion. After some discussions with one of the pastors about his ministry school students' fears of science, I agreed to begin teaching some short courses.

"Thank you, Ruth, for your question, and I think we can address this idea of a war here."

I went on to describe the origins of this idea at the end of the nineteenth century. The book most often referenced in support of there being a war was published by Andrew D. White (1832–1918) in 1896, *A History of the Warfare of Science with Theology in Christendom*.[1] White's title was misleading to many, however. Rather than concluding that science was in opposition to theism, he wrote of his conviction that someday, after the defeat of dogmatic theology, "Science . . . will go hand in hand with Religion." He believed that there was a "Power in the universe, not ourselves, which makes for righteousness." White's war was to free religion from dogmatism using science. He was supported in this by many scientists in the early twentieth century. For example, 1915 Nobel laureate William Lawrence Bragg (1890–1971) described opposition between religion and science as the opposition between his thumb and fingers, continuing, "It is an opposition by means of which anything can be grasped."[2]

"We need to start with White's book," I droned on when a young woman named Abigail interrupted me in her British accent.

---

1. White, *Warfare of Science with Theology*.
2. Bragg, *World of Sound*.

"But I'm not interested in Ruth's 6,000-year-old earth. That is an American thing that we in Europe have long ago gotten over. In Britain, the real issue is the 'New Atheism' being promoted by Richard Dawkins at Oxford University. He is supposed to be a 'professor of public understanding of science,' but many of his Oxford colleagues argue that he is wrong about science and religion. I want to know if he and the Four Horsemen of Atheism are right."

Several students immediately began speaking, making an uproar. The teacher in the next room looked in to see what was going on. As the students began arguing with each other, I realized that my teaching plan for gradually drawing out the history of science and religion wasn't the best approach. While the so-called young-earth argument that Ruth had asked about would have to be discussed because most of my students were American, addressing the claims of the New Atheists appeared to be a better way to begin. However, some of the foreign students appeared more knowledgeable than me, and I knew I would need to do some more reading.

In my next lecture, I contrasted two of the worldviews described by James Sire,[3] Christian theism and atheistic naturalism.

I explained that "the assumptions of atheistic naturalists imply that the universe is completely deterministic, a closed system. In contrast, the assumptions of Christian theists imply that the universe is not deterministic, but an open system. The distinction is crucial because a closed system implies that there is no God and everything we see is the product of deterministic processes. In contrast, an open system allows for God as the creator of all we see."

As pointed out by Sire, I explained, people can and do choose between these two worldviews because it is not possible to prove from first principles that the universe is either closed (naturalism) or open (theism). However, in his 1986 book *The Blind Watchmaker* Richard Dawkins (1941–) claimed that evolutionary theory is capable of completely explaining the complexity of life on a deterministic basis. Based on that, he argued, the universe is likely deterministic; no designer is needed.

I continued a seemingly slow lecture, wanting to stay ahead of my students. "Beginning with his 2004 book, *The End of Faith*,[4] Sam Harris has defined the terms of reference for the New Atheists, and subsequent

---

3. Sire, *Universe Next Door*, ch. 4.
4. Harris, *End of Faith*.

books by Dawkins, Daniel Dennett (1942–), and Christopher Hitchens (1949–2011) have elaborated the major tenets of their beliefs."

Struggling to come to a point, I explained: "The openness with which they express their disbelief in the supernatural is amplified by their high level of confidence and often bombastic style. They would like to put themselves forward as representing 'a body of established opinion widely accepted as authoritative.'"[5]

"So, in answer to your question from the last lecture, Abigail," I continued, "the arguments of Dawkins and Harris and a few others are complicated and have been heavily criticized both theologically and scientifically."

"Okay, okay," she responded vehemently, "I get it that this is all complicated, but what do you say? Church leaders seem to pussyfoot around this rather than responding directly."

I was taken aback by her anger and tried to elaborate.

"Sorry. I understand your frustration. I don't agree with Dawkins and his colleagues, but you knowing my beliefs is insufficient. You must develop your own understanding. The Christian philosopher Alvin Plantinga's book *Where the Conflict Really Lies*[6] is a good starting place. He argues there that Dawkins and his colleagues' arguments are not strong and far from convincing logically. Instead of evolutionary theory disproving religion, as they claim, Plantinga sees a 'deep concord between science and theistic religion.'"

I had the class's attention now because they had wanted someone to tell them what to believe. But I couldn't leave it there. I continued, "Plantinga describes several meanings for the term evolution as a way of describing what Dawkins means by evolutionary theory. First is that life 'developed from non-living matter without any special creative activity of God': naturalistic origins. Second, and connected, is that the diversity of life developed from the initial simple life forms that we observe in ancient rocks into today's complexity by descent with modification. And finally, that these changes have arisen by random (that is unguided) natural selection of random genetic variability. The wrinkle, of course, is the unguided part."

I went on, following Dawkins's argument that although the likelihood of the genetic variation we see is very small, it is not "astronomically improbable."

5. Plantinga, *Where the Conflict Really Lies*.
6. Ibid.

"That is to say," I continued, "unguided evolution is a plausible if unlikely mechanism to explain today's diversity. However, as Plantinga points out, Dawkins's conclusion that this is plausible doesn't make it necessarily true, and hence not a reliable basis for choosing determinism over theism."

Plantinga goes on to counter additional arguments of Dawkins and his colleagues. For example, Dennett in *Darwin's Dangerous Idea* argues that there is meaning in life, but that its origin is in matter itself, nothing outside of the material world is required. Dennett suggests that "An impersonal, unreflective, robotic, mindless little scrap of molecular machinery is the ultimate basis of all the agency, and hence meaning, and hence consciousness, in the universe."[7]

"But," I noted, "many people have objected to this idea, for example, John Locke in 1689 noted the unlikeliness that 'ever pure incogitative Matter should produce a thinking intelligent Being.'[8] This is a more powerful argument against the atheism that the Four Horsemen have advocated and is what I base my rejection of their arguments on."

Abigail had not objected to my insisting on guiding the class through Plantinga's book, and even led a round of applause at the end of the lecture. I had survived presenting the Four Horsemen, but of course still had to address Ruth's 6,000-years question.

The War within Christianity

"I have to say up front that the so-called war between science and religion pushed me away from Christianity and towards atheism as a child. I do not believe that the world is only a few thousand years old. And I object to those who teach this; there is no convincing evidence for this."

Having learned from my previous experience, I began this next lecture more aggressively, directly stating my conclusions. The class, even Ruth who initially broached this question, stood and applauded. Where I am always anxious to approach controversial topics slowly with a carefully justified argument, the direct approach seemed to work better. That is, it would work better if the students could be led to develop their own understanding.

Continuing my lecture, I traced this idea of war back to Andrew White's 1896 book *A History of the Warfare of Science with Theology in*

---

7. Dennett, *Darwin's Dangerous Idea*, 203.
8. Locke, *Essay Concerning Human Understanding*, bk. 4, ¶10.

Christendom. White's focus was clearly on conflicts among Christians rather than with atheists. He saw the problem being religious dogmatism, and various forms of such dogmatism have developed. Ruth's experience with this conflict has proven common among students here, and I was reminded of just how common a few days later when a tall, intense man with sandy hair and horn-rimmed glasses approached me in the church lobby.

"Are you Dr. Smith?" he asked.

I didn't recognize him but saw that he was holding two books. "Yes," I said, hesitantly. Few here knew I had a PhD in science, and being addressed with that title put me on alert.

"Well, I heard that you were teaching about the age of the earth," he continued, edging closer. "I thought you should have these two books. They will help you understand the truth."

He hadn't said how he knew what I was teaching, and I wondered if there was a student who wasn't so pleased with my assertions. I took his books, and I immediately recognized one as being Henry Morris's (1918–2006) *Science and the Bible*.[9] That book champions the idea that the earth is only a few thousand years old. The other one was new to me.

After lunch that Sunday I opened the man's second book, Marshall Hall's (1931–2013) *The Earth Is Not Moving*.[10] I realized that I had not asked the man's name, trying to distance myself from him, I guessed. I quickly saw that perhaps I should distance myself from this book as well. The essence of Marshall Hall's argument was that the Bible describes the earth as not moving and the sun as rotating around the earth. This was important because he claimed that the Bible is "the source of all Truth."

I quickly began to tire of Hall's ranting style and paused to look his book up and see what others had to say. Unsurprisingly, Hall's opponents drew on familiar cosmological observations and on less well-known (to me) biblical observations.

When I began my next lecture, I focused on Henry Morris's book, leaving Marshall Hall's aside. Morris seems to have had a big influence on some Christian groups that focus on using science to support literal interpretations of the Bible. He begins with the claim: "One of the most amazing evidences of the divine inspiration of the Bible is its scientific accuracy." In his first chapter, he listed biblical observations that he claimed would only later be shown by science to be true. These included: Jeremiah's contention

9. Morris, *Science and the Bible*.
10. Hall, *Earth Is Not Moving*.

that the stars cannot be numbered (33:22), Isaiah's description of a circular earth (40:22), descriptions of the water cycle by the author of Job (36:27–29) and also by King Solomon (970–931 BC) in Ecclesiastes 1:6–7, and the biological significance of blood recorded by Moses in Leviticus 17:11.

Morris recognized that he was interpreting these biblical passages, but he didn't acknowledge that he was also interpreting other passages by omission. For example, Daniel's dream of a tree that could be seen from all the corners of the earth (Dan 4:10–11), an important observation for Marshall Hall's thesis that the earth is not moving, got no mention by Morris. Also, David's observation in Psalm 93:1 that "the world is established; it shall never be moved," which Hall claimed was evidence that the earth is not moving, got no mention.

Although I hadn't found much of help in the sandy-haired man's two books, I recognized many questions that had proven stumbling blocks to me growing up, questions that had proven difficult for many of Ruth's classmates now. Those questions reminded me that although I was conversant with the main conclusions of modern science, especially astronomy, geology, and biology, I also needed to delve deeper into the various interpretations of many biblical passages.

My previous focus in reading the Bible had been on understanding the history of the Hebrew people in the Old Testament and its relationship to Jesus' message in the New Testament. I had not really thought of the Bible from a scientific perspective. Of course, I was familiar with what looked like two creation stories in Genesis and references in a few other passages. For example, David also described creation, albeit more poetically in including God creating the heavens using the clouds for a chariot riding "on the wings of the wind" (Ps 104). Such text is hard to take literally, just as are the creation stories of many other cultures, which seemingly must be taken poetically or figuratively.

But the intense scrutiny of such biblical texts to support the present-day observations of nature is not unique to Hall and Morris. Many others also offer interpretations of biblical texts as reliable sources of scientific information. If it is, then interpretations of biblical text must be consistent across the Bible and with what we observe about nature.

Not everyone agrees that the Bible should be interpreted in this manner. For example, one Christian group, dubbed "theistic evolutionists," holds that "Genesis is not meant to teach what they would see as scientific information."[11] Another group, old-earth creationists, has focused on the importance of carefully considering our interpretations of observations of nature and of observations and interpretations of biblical text because these interpretations must be consistent with themselves and with our observations of the world.[12] Finally, a third group, young-earth creationists, takes a different approach, arguing that the observations of science must be reinterpreted based on a literal interpretation of the creation story in Genesis.[13]

These specific arguments are current points of contention within Christianity in its millennium-long history of interpreting science, and they must be seen in a broad historical and theological context.[14] The often-controversial arguments within Christianity among these three and other related perspectives about the interpretation of our observations of the world and our observations of the Bible have proven less than helpful. Even as I was confused and pushed away from Christianity by these controversies as a child, young people continue to struggle with them today.

This supposed war between religion and science was only the first of several topics that I would address with the students at the school of ministry. Other topics regarding human nature came to fascinate me as well, including our cognitive capabilities, the human genome, and the nature of freedom. My father's voice had, up to this point, been influential in my understanding of human nature and would be challenging to silence, but it fueled me to obtain a deeper understanding.

---

11. See http://www.biologos.org.

12. See http://www.reasons.org.

13. See http://www.answersingenesis.org.

14. See Collins, *Science and Faith*; Pearcey and Thaxton, *Soul of Science*; Brown, *Abacus and the Cross*.

12

Being Human

"Sit on that downed tree," Pop said. "I'll drop down the hill aways and then make a drive back toward this meadow."

I settled in on the maple tree, the moss wetter than I expected, and waited. After about ten minutes I began to hear Pop crashing up the hill through the undergrowth; then silence. That was his favorite trick, making a deer move by suddenly becoming quiet. I calmed myself.

A few minutes later I saw the beautiful forked horn deer emerge into the meadow. My seven-millimeter saddle carbine lifted and nestled into my shoulder with its sights lined up on the buck's heart as if this was a normal thing rather than my first kill. All I had to do was touch the trigger.

I was calm until flames spat from the short barrel and the deer silently dropped, blood spurting out of its chest. Then the report of the carbine echoed back down the valley, and suddenly my hands and heart began shaking.

"Good shot, Tim," Pop called out as he entered the meadow and found the deer. "Right through his heart."

We stayed motionless for a while, waiting to see if another deer would appear through all this commotion. My shaking began to subside, and I was glad there was no second deer. I wondered if Pop had killed any Germans or Japanese during the war and if he had shaken as I had. I wanted to ask him but didn't; some things are too personal, at least within our family. I was sure that killing a deer would be very different from killing a person because, even though all living creatures are formed of the same DNA, there is something unique about humans. Killing a deer is much simpler than what soldiers face.

I didn't really understand the significance of my question that day until many years later when I read evolutionary biologist Simon Conway Morris's description of a human as a "sentient being that sees meaning."[1] Realizing that we see and even insist on finding meaning in our lives means that killing a human denies that person the opportunity to work out his purpose, his meaning.

But what is the biological difference between that deer and me? We share life and death and our general mammalian physiology, down to the very similar protein synthesis parts of our DNA. The physical similarity of man to some other organisms, notably primates, is evident, and by the 1860s people were speculating which of the great apes and chimpanzees were our ancestors. This was captured in a drawing showing a row of primates, from monkeys on the right to increasingly erect humanoid forms on the left. The implication that these primates descended from one another was obvious but wrong. For example, analysis of DNA suggests that if there were a relationship, it would have been about six million years ago, millions of years before the oldest *Homo sapiens* fossils ever found. Any possible relationship would not be one of a line of descent, but rather of a remotely shared ancestor.

The genetic codes of chimpanzees and humans are nearly identical, for example, 99 percent identical in the portions of the DNA that control the essential protein synthesis in the cells. The other portions of the DNA are more diverse, but still remarkably similar, with humans, mice, cows, and chimpanzees all between 84 percent and 90 percent similar.[2] These results suggest that protein synthesis is similar among mammals, even while morphology ranges widely. But these similarities are overshadowed by the radically different cognitive faculties of humans. However we came to be, and whatever similarities there are to other animals, we think in ways that no other animal does.

What Is Human?

Growing up in our remote valley of Camp Creek offered me few contacts with other races, although the valley was named for the remnants of Native American camps along the creek where my brother and I picked up

---

1. Conway Morris, "Evolution and Convergence."
2. See *National Geographic* at http://ngm.nationalgeographic.com/2013/0/125-explore shared-genes.

arrowheads. Thus in 1965, as I was graduating from high school, I was totally unable to comprehend the race riots in Watts, a suburb of Los Angeles. My father had held "blacks" (African Americans) and "Okies" (people from Oklahoma) in the same low esteem. The latter had moved to Oregon during the Depression, and he resented them for competing with him for jobs.

"Blacks and Okies were okay," Pop would say, "but they are just not our kind."

As for me? I thought whatever Pop thought. I remember finding anthropologist Carleton Coon's (1904–1981) 1962 book *The Origin of Races*[3] in the university library. Coon offered an explanation of those riots, and I was unprepared to question its validity. His argument was that humans had evolved separately and sequentially five times from *Homo erectus*, forming five separate races. Coon's Caucasoid race was hundreds of thousands of years older than the other races, and hence, progressive evolution would imply that had developed to be wiser and more civilized. The other races, especially the Congoid race, were less advanced. It all seemed consistent, Coon's theory and Watts's experience and Pop's thinking. In retrospect, my gullibility to bad science is embarrassing.

The regional differences in language and appearance of humans were early used to categorize people groups, as evidenced in the long history of tribal and ethnic conflicts and of groups enslaving one another. These differences were the origins of the concept of race that developed beginning in the seventeenth century. By the time of Darwin's voyage aboard the HMS. Beagle in the 1830s, the nature of the many people groups that he and other travelers were encountering was being widely discussed. Anthropologists, biologists, statisticians, and others were drawn into the task of describing the various groups using morphological measurements. Collecting and studying human skulls was popular, with cranial capacity and skull shape being regularly measured. In addition, theories about the meaning of bumps on the surface of skulls, phrenology, were well developed and drawn in to describe the groups of people being observed and measured by biologists.

The critical point was not the existence of unique people groups, which were obviously different in some ways from one another. The point was the interpretation of these differences. Genesis provided one grounding point, namely that God created man once, not multiple times, an interpretation that Darwin, in his theory of human evolution, supported long

---

3. Coon, *Origin of Races*.

after he himself had abandoned Genesis. The other interpretation, contrary to Genesis, was that these many groups had originated separately at different times. The importance of this argument was that it allowed the different groups to be interpreted as different species and assigned different values.

Darwin had anticipated a strong challenge to his theory because of this theory of multiple origins. Some scientists and preachers argued that different people groups had been created and evolved separately, and hence should not be expected to be equal. Even though the African slave trade had been outlawed in England in the late 1700s, slavery was still being phased out in its colonies. Further, in America, the regional differences in slavery were rising toward the breaking point of the U.S. Civil War. His publication of his theory drew him into controversy, something he did not like.[4]

Moses' description of the origin of people included both specific and general geography. Specifically, Moses (Gen 2:14) described the garden of Eden as at the headwaters of four rivers, two of which were the Tigress and the Euphrates. The oldest fossilized humans, *Homo sapiens*, found so far are in Central Africa, in Kibish, Ethiopia, in sites dated to 200,000 years ago.[5] More generally, Moses describes people congregating in one area, Babel, being "dispersed over the face of the whole earth," and ultimately being separated because God "confused" their language (Gen 11:4). Evidence for such dispersal is seen in the discovery of fossilized humans in other parts of the world dating to roughly 100,000 years ago in South Africa, China, and Israel and to roughly 35,000 years ago in Europe. Further, there are several thousand languages spoken, almost all of them quite complex. Humans appear to have expanded into virtually all parts of the earth with relatively minor genetic differences and some obvious morphological and linguistic differences, such as skin color and facial shape.

Biological analysis of the genetic differences has helped us better understand human origins. For example, spatial gradients have been detected using genetic analysis methods called microsatellites. These genetic markers for samples from modern-day individuals chosen randomly over the earth have shown gradients both within and between continents.[6]

These spatial gradients are consistent with the results of other, more specific genetic analyses of two small specialized DNA segments. One is transmitted from mother to daughter and the other from father to son.

---

4. Desmond and Moore, *Darwin's Sacred Cause*, 342.
5. McDougall et al., "Stratigraphic Placement."
6. Serre and Paabo, "Evidence for Gradients."

The first type of segment is found in microscopic organelles called mitochondria. These tiny structures are essential to life, being responsible for converting food to energy. Only the genes in the mothers' mitochondria are passed to the next generation, with no mixing with the genes in the fathers' mitochondria. Thus, the only changes in mitochondria are due to mutation, not due to recombination between the sexes. Hence, any such changes occur much more slowly than in other genes. The second DNA segment is the Y-chromosome, found only in males. This segment contains 50–60 genes and passes mostly intact (without recombination) from fathers to sons. Analysis of various sets of mitochondrial DNA and Y-chromosome genetic data, in conjunction with anthropological and archaeological data, have suggested that over the last 200,000 years *Homo sapiens* have expanded out of Africa, east into Asia, south into Australia, north into Europe, and then west into North and South America.

The genetic differences suggest that even though humans are all one species, there are global and regional differences in the frequency of some genes. These have been described for skin color, face shape, and disease susceptibility. For example, the long-recognized association between malaria susceptibility and the sickle cell anemia trait has recently been shown to reflect microevolution, suggesting new approaches to developing an antimalarial drug.

Genetic Determinism

Although African Americans did not live in Camp Creek Valley when I was growing up, immigrants from Oklahoma did. Pop was fond of pointing out ramshackle homes, all of which he said were occupied by Okies. Their extended families occupying neighboring no-longer-mobile homes were the recipients of his scorn.

The origin of Pop's attitudes may have been that they might compete with him for jobs, but the tolerance for his attitudes dates back much further. Darwin's, Wallace's, and Spencer's theories involving the survival of the fittest raised the question of how much of our behavior, and thus our survivability, was determined genetically. For example, in 1880 E. Ray Lankester (1847–1929) interpreted the Darwin-Wallace theory in his study of parasites as examples of what he called "degenerate evolution." He thought, "Any new set of conditions which render a species' food and safety very easily obtained, seems to lead to degeneration."[7]

While examples of such degenerate species were easy to come by, Lankester also projected his theory of degeneracy beyond biology and onto human societies. Similarly, Spencer used the Darwin-Wallace theory's "survival of the fittest" as an analogy about human society, a theory called "social Darwinism." That is the philosophy that the strong and capable in society should prosper while the weak are doomed to die off. Helping them would only weaken society.

The extent to which human behavior is determined by genetics, like eye color or facial structure, became the target of Spencer's social Darwinism theory. Just as animal husbandry had improved farm animals by eliminating individuals with less desirable traits, perhaps the behavior of populations could be improved in the same way. The crucial assumption for this theory was that human behavior was substantially determined by genetics, that, as my father had said, people could not change. It is from here that one of the greatest science-based evils of the twentieth century emerged.

The origins of that evil began in the first decades of the twentieth century, when many well-known people supported what became known as the eugenics movement, for instance, Alexander Graham Bell, Winston Churchill, John Maynard Keynes, and Woodrow Wilson.[8] By 1910 Charles B. Davenport (1866–1944) had added the study of human inheritance to his Carnegie Foundation–supported Biological Experiment Station for the study of evolution. Located at the Cold Springs Harbor Laboratory in New York, the Eugenics Records Office collected family pedigrees, following the methods of selective breeding of racing horses and show dogs. Davenport's interest, however, was not about the best-of-breed, but rather those families who appeared to be far from that.

The support for Davenport's eugenics research reflects widespread approval of these ideas, even though there was criticism from some biologists. For example, geneticist Thomas Morgan (1866–1945) thought that eugenicists didn't actually understand the traits they were identifying in their family pedigrees. The heritability of the hypothesized traits was central to the theory, but neither care in definitions of traits nor evidence for

---

7. Lankester, *Degeneration*.
8. Farber, "U.S. Scientists' Role."

their heritability was a priority. One example is a trait termed "sea-lust," or dressed up, thalassophilia, which Davenport described as:

> One of the most striking characteristics of sea-lust is that it is wholly a male character ... so the appeal of the sea develops under the secretion of the germ gland in the boy. It is theoretically possible that some mothers are heterozygous for love of the sea, so that when married to a thalassophilic man half of their children will show sea-lust and half will not.[9]

Even such obviously poor definitions of traits failed to draw the attention of most biologists and support for eugenics continued even after the Carnegie Foundation's support for Davenport ended in 1939. By that time some 60,000 forced sterilizations had been performed in the United States, progress toward the goal announced by the American Breeders Associations of sterilizing 10 percent of the United States population.[10]

But beyond the biologists and the state governments who participated in these sterilizations, either by their silence or by their advocacy, others had been listening. One group was the German Nazi Party in the 1930s, which eventually performed at least 360,000 sterilizations of "undesirables" by one definition or another.[11]

However, even following the Nazi era, these ideas persisted. For example, Julian Huxley continued to advocate eugenics ideas even in his 1962 *Modern Synthesis*, coupling it with another suspect assumption, namely, the "Evolutionary Progress" of his last chapter. From these two ideas, he recommended that human progress could be made by "separating the two functions of sex—love and reproduction—and using the gametes from a few highly endowed males to sire all the next generation ... then all kinds of new possibilities would emerge."[12] He saw possibilities for human progress through the development of "true castes," drawing on the biology of cooperative social insects that would be selected for an "increase in intraspecific co-operation."

The closing of the Cold Spring Harbor Laboratory's eugenics work was the end of the family trees that had been used to implement eugenic thinking in the United States. But some, like Julian Huxley, had continued to write widely about methods of implementing eugenic improvement of

9. Davenport and Scudder, "Naval Officers."
10. Kühl, *Nazi Connection*.
11. Sofair and Kaldjian, "Eugenic Sterilization."
12 J. Huxley, *Evolution*, 573.

the human genome. The use of these ideas has recently been traced through justification for President Nixon's 1972 welfare reform and into the present human genome era.[13]

The inability or unwillingness of biologists and anthropologists and of theologians and philosophers to identify and test key assumptions of racial theory and genetic determinism is concerning. This failure condemned many to sterilization in the twentieth century and challenged the assumption that although science can get things wrong, it is sufficiently self-correcting and such errors are acceptable. Sometimes, perhaps, but in this case, the science was fundamentally flawed and self-correcting only on the scale of human generations. Further, while new techniques have revealed many of the assumptions of the eugenics movement to be erroneous, the ideas developed then still have the potential to guide the implementation of human genomics in the twenty-first century.

Human Cognitive Faculties

Eventually explaining the existence of our cognitive faculties would split Darwin and Wallace. They both agreed "that natural selection cannot develop an organ beyond the needs of its possessor." They also both realized that our human cognitive faculties are just that, beyond our simple survival needs. To Darwin's chagrin, however, Wallace took this idea to its logical conclusion:

> We must therefore admit the possibility, that in the development of the human race, a Higher Intelligence has guided the same laws for nobler ends. Such, we believe, is the direction in which we shall find the true reconciliation of Science with Theology on this most momentous problem.[14]

In contrast, Darwin published another book in 1871 where he focused on the application of his theory to humans, *The Descent of Man and Selection in Relation of Sex*.[15] There he argued that humans had evolved from lower animals through the increased need for greater intelligence raised by the intricacies of mate selection, for example, the elaborate and unlikely tail feathers of male peacocks. Interestingly, Protestant minister Antoinette

---

13. Lombardo, *Century of Eugenics*.
14. Wallace, "Geological Climates," 394.
15. Darwin, *Descent of Man*.

Blackwell (1825–1921) criticized his explanation as missing out on the role of females, being based on the "assumption that the male is the normal type of his species."[16]

Darwin, however, had not entirely convinced himself that human faculties could have developed from random natural selection. He later wrote to a friend:

> With me the horrid doubt always arises whether the convictions of man's mind, which has been developed from the mind of the lower animals, are of any value or at all trustworthy. Would any one trust in the convictions of a monkey's mind, if there are any convictions in such a mind?[17]

While biologists have no explanation beyond random natural selection for our cognitive faculties, Scripture does. Moses used the Hebrew word *tselm* in describing God's creation of man, meaning likeness or resemblance. The nature of likeness becomes clearer in Genesis 2. Here, Moses augments the name of God, *elohiym*, with *Yahweh,* hence *Yahweh elohiym,* the Lord God. *Yahweh* is a personal name that God used in describing relationships with humans (Exod 3:13–5). In the more elaborate description of the origin of man in Genesis 2, the Lord God "breathed into his nostrils the breath of life." Thus, God seems to be entering into an intimate, direct relationship with Adam. God then formed woman from Adam's body and walked with them in the garden that he created for them. Further, Moses described God giving humans both responsibility—to care for the garden of Eden—and freedom to choose, especially to refrain from eating from one specific tree in the Garden.

Wallace and Moses explained the origin of our faculties in our being created by something outside of nature, beyond the argument of natural selection. Similarly, Christian philosopher Alvin Plantinga and atheist philosopher Thomas Nagel (1937–) addressed the origin of our cognitive faculties recently in two separate books.[18] They also each wrote reviews of each other's book.[19] Both philosophers concluded that Wallace was correct and Darwin wrong. Further, they agreed that Neo-Darwinism had introduced nothing new to explain what we are, our capabilities.

16. Blackwell, *Sexes Throughout Nature.*
17. Darwin, *Darwin Correspondence Project*, letter to William Graham, July 3, 1881.
18. Plantinga, *Where the Conflict Really Lies*; Nagle, *Mind and Cosmos.*
19. Plantinga, "Why Darwinist Materialism Is Wrong"; Nagel, "Philosopher Defends Religion."

Plantinga went on to argue that there is concord rather than conflict between science and Christianity. This concord is due to our having been created in the image of God, *imago Dei*, and thereby receiving "knowledge of ourselves and of our world." While accepting the failure of Neo-Darwinism, Nagle, in contrast, states that he lacks Plantinga's *sensus divinitatis* and further confesses to being strongly averse to the idea that he, too, could experience it. Unfortunately, Nagle has no alternative explanation for our cognitive faculties, while Plantinga goes on to suggest how to test his idea of concord between science and Christianity. We could resolve this issue, he argues, by examining our cognitive faculties to see if they are consistent with the idea that they arose from our being created in the image of God.

Being created in the image of God, of course, does not mean that we physically look like God. That was a device that Michelangelo used in his Sistine Chapel paintings of God creating both the earth and Adam and Eve. In the absence of a biblical physical image of God, Michelangelo portrayed the events of creation using an image based on our own appearance. Focusing there, however, could lead us to miss the more significant spiritual implication: "The Spirit of God has made me, and the breath of the Almighty gives me life" (Job 33:4). *Imago Dei* is not about opposable thumbs and enlarged cerebral cortexes, but rather about something beyond the physical, what Moses described at the beginning of Genesis, the Spirit of God hovering over the waters, and over us.

Abstract Reasoning

Abilities that are unique to humans include abstract reasoning, self-awareness, and a need to know meaning. Abstract reasoning is our most tangible cognitive faculty and has allowed us to develop logic and mathematics, art and music, and the ability to ask questions like: "Does the world embody beautiful ideas?"[20] This has also allowed us to understand and manipulate our world and the universe in ways that are completely beyond that of any other organism. The usual success of these undertakings is ample evidence that our abstract reasoning allows us to accurately understand how the physical world works. Further, as a social organism, we have been able to far exceed the cooperative capabilities of other social organisms, for example, reef-building corals, web-building spiders, and colony-building

20. Wilczek, *Beautiful Question*.

insects. And we've demonstrated a much greater capacity to wage war than any war-like ants.

Our invention of mathematics gives us the capability of measurement. That allows us to observe what is around us in a repeatable manner, providing a basis for quantitative activities such as trade and building. But abstract reasoning also leads to the development of logic and, along with mathematics, to the development of science. Some historians of science have argued that science has developed "largely by debates *among Christians* over which philosophy of nature gives the best way to conceptualize the kind of world God created and the nature of his relationship to it."[21] Science arose within Christianity because it alone of all cultures encourages us to seek God in his creation (Rom 1:19). Further, as Loren Eisley wrote, Christianity provided the basis for science: "It is the Christian world which finally gave birth in a clear, articulate fashion to the experimental method of science itself."[22] Nancy Pearcey and Charles Thaxton lay out the intimate connection between Christianity and science, agreeing with Plantinga and Nagle that the apparent conflict between science and religion arising in the twentieth century is false.

Self-Awareness

I remember discussing with my brother Eric that there was something in me that could choose to look out at the world or choose to look inward. Looking out was visual, like Superman's X-ray vision, a beam going out from my eyes. Looking inward was very different and scary because I could see but couldn't touch what was in there, inside of me. Eric may have led me to explore this aspect of myself prematurely, but there was no going back. He assured me that what I could see looking inward was made of the same stuff as what was outside of me. I remember a discussion of skin, and that without our skin we would be able to see inside ourselves. But we weren't talking about things you could visually see, even with X-ray vision.

Self-consciousness, instead, was apparent in Moses' description of the first humans, initially when Adam responded to the creation of Eve with the phrase, "This at last is bone of my bones" (Gen 2:23). He was aware that he had been waiting, "at last," and that Eve was both part of him and separate from him. Later, after exercising their freedom to choose and choosing

21. Pearcey and Thaxton, *Soul of Science*.
22. Eisley, *Darwin's Century*.

against God, Moses wrote, "Then the eyes of both were opened." They could then see into themselves, their failure and their shame.

This self-consciousness leads to ethical reasoning, and we spend a lot of energy and time discussing right, wrong, and freedom. People across all cultures appear to share an underlying understanding of right and wrong.[23] Ethical reasoning is not simple, and we learn to weigh the specifics of a situation. But in the end, regardless of how we exercise our freedom, we are drawn to evaluate the ethics of the responses of ourselves and of others.

Ethical reasoning was also apparent in Moses' description that Adam and Eve, through their opened eyes, "knew that they were naked," and they knew shame and that they were guilty of choosing against God. Man's responsibility for taking care of the Garden and his freedom to choose underlays Moses's description of God's relationship with his people. Many more ethical issues soon arose—issues that were addressed through God establishing laws, such as the Ten Commandments, and importantly, establishing a covenant relationship involving the repentance for breaking those laws through sacrifice. People later also experienced a new covenant, the sacrifice of God's Son, and grace in lieu of their continued sacrifice. These were concepts that required additional ethical reasoning, reasoning that we continue to develop today.

Inherent in ethical questions is our response to the needs of others. One puzzling behavior of people is helping others at the expense of themselves, altruism. Darwin and Wallace's suggested that the fittest are those who take care of themselves and their offspring, maximizing the number of their genes in future populations. But while people do care for their families, they also sometimes care for unrelated people.

Altruism also reflects feelings of empathy, the feeling that you understand and share another person's experiences and emotions. In August of 1976, I understood empathy through stone when I first encountered Gustave Vigeland's statues in Oslo, Norway. The old man and the old woman he cradled were sharing something from their life together, and I felt it empathetically, even though I did not know what it was. Vigeland's stone showed me the power of empathy.

The prophet Jeremiah knew the difference between empathy and compassion when he described his anguish: "Oh that my head were waters, and my eyes a fountain of tears, that I might weep day and night for the slain of the daughter of my people" (Jer 9:1). St. John later described Jesus'

---

23. Lewis, *Surprised by Joy*, appendix.

instructions on how we ought to respond to others: "That you love one another" (John 13:34), and further, that love has no limits: "Greater love has no one than this, that someone lay down his life for his friends" (John 15:13).

There is something unique about humans but putting our finger on exactly what that is has proven difficult. One aspect is that our cognitive abilities allow us to be sentient and another is that we see meaning in our lives. Our cognitive abilities also allow us abstract reasoning and to be self-aware of our responses to our environment. Despite our uniqueness, however, we are also beings whose responses to the environment are to some degree determined by our individual genomes. Being human is a complex business, and there are many ways in which we are distorted, both by our genome and by the events of life. As much as that makes us sound like determined beings unable to change, I've discovered that the effects of our environment can be healed, which leads to change.

13

Being Healed: 2011–2018

THE FILM IN MY mind's eye always started with the little boy sitting quietly on the hillside in front of his home, with his trucks and his B-29 Superfortress bomber. Then he would start playing with his toys, his airplane dropping invisible bombs on the trucks and the people in the trucks flying everywhere. Each time the film began, I knew it would continue for a minute or so. I was transfixed and watched it over and over again, but not feeling anything. The film would begin spontaneously, sometimes at inopportune times.

One inopportune time was when I began a counseling training class in our new church. We had been doing some pastoral counseling back in Cape Cod before we retired, so I signed up for this class, hoping it would help me channel the feelings of compassion that continued to bring up tears.

During the first class, the teacher asked for a volunteer, so she could demonstrate her counseling approach. I didn't raise my hand, but no one else did either. I was slightly intimidated by this woman, too much authority, too much control. Sighing, she looked around and then turned to me: "Tim, would you be my volunteer?"

I was there to learn how people here counsel others, not to be counseled myself. I felt anger rise up at her for using her authority to control me. I felt bullied, but I didn't want to appear uncooperative. She seated me in the middle of the room and asked me to close my eyes. Unwillingly, I did, reassuring myself that I could stonewall her. But instantly the film began again, the little boy on the hillside. This time I felt that little boy's anger rise up. I remembered not being allowed to begin school when I had expected because the new superintendent changed the rules. I remembered that my parents had not intervened. I remembered sitting there in the dirt, angry, so angry. And I couldn't stonewall this counselor, suddenly sobbing.

She ended the demonstration abruptly, not having expected my response, and we met later, privately. I tried to explain what had happened, but I didn't really know. Since our first public counseling session, I had remembered other times when anger just rose up, out of my control. Once during recess early in the fourth grade, I had felt that anger as I continued to pummel the school bully long after I had made my point—until I was satisfied with enough blood. Though I explained what I saw anger do, the school principal did not allow me to draw a distinction between me and that anger. That day I knew there was something more inside me than physical stuff, something the school principal insisted I needed to control. I remembered another day, this one much later back in Cape Cod, when I lost control after the store owner had refused to listen to my more than reasonable request. My good friend had forcefully extracted me from that store.

The counselor didn't ask me what I felt about all of that but rather suggested I ask the Holy Spirit to help. Strange things happened that day, but over the next few weeks the vision kept coming up, and anger with it. During one meeting with the counselor, the vision changed. This time, it began as it normally did, but then another figure showed up. I saw Jesus walk over to the little boy, squat down to inspect his toys, ask him a question that I could not hear, and then take his hand and walk away with him. I lost it all over again, sobbing until I couldn't sob anymore. In an instant I felt something heavy lift off me—then I felt joy. Then I felt peace. I could still visualize the scene in front of my home, but the little boy wasn't there, and suddenly I felt no anger towards the school superintendent.

The film never came back, and gradually the details of the scene faded. The sharp regrets over my many conflicts with authority lost their edges. I couldn't find any anger towards the fire marshal in the boiler room and the teacher that allowed us to be there. Other authorities in the various schools and in my career once were bitter memories that now began to fade as well. I recalled my resentment when I had left questions unasked, and my anger when I had contradicted authority figures and asked those questions anyway; those resentments began to fade away.

Looking back, I realized that conflicts with authorities had been significant in my life. Sometimes acquiescing to and sometimes contradicting authorities had both taken their toll on me, and that toll was like a debt that could not be paid off. It continued to subtract from me. I can remember the direct effects of many of those authorities, obviously the original fire marshal and the grade school teacher that allowed the Boiler Room Boys in the first place. I can remember the high school geography teacher who intimidated me to not ask about continental drift and the guidance counselor's refusal of my request to study Russian history.

There were more incidences when I backed away from the questions I wanted to ask. For example, the science fair teacher encouraging me to lie and my college professor controlling my fieldwork in the South Pacific. But strangely, after Jesus walked away with the little boy on the hillside, I no longer felt intimidated by my fire marshals, no longer afraid to pursue some questions because I might get exposed. I also lost my fear of the religious authorities: the Sunday school teachers, the pastor advising not to pursue sensitive questions, and the book writers claiming to have all the answers about science or about religion or about both.

Answering God's question and answering the counselor's invitation had begun a process of healing. Through that, my continuing asking questions about what is true was replaced by the courage of truth-telling. That changed me, something that one fire marshal, my father, had claimed could not occur. It is hard to write that your father was wrong, but he was.

Believing in Healing

After the little boy walked away, I began to consider the healing events that Margene had experienced back on Cape Cod. I had discounted the initial healing of her vision because I couldn't actually see what had happened, although Gayle said she heard God saying that this is a down payment on the healing that is to come. The healing of her extreme fatigue just before we went to the conference in Connecticut was evident, but still not something I could prove beyond her testimony. I was surprised at how I had been willing to doubt her in order to preserve my belief that people don't spontaneously heal, or, indeed, even change. Besides, her health had continued to decline, especially her sense of balance.

Although she was not using a wheelchair, she frequently steadied herself by using walls. The first afternoon of yet another healing conference, someone felt that we should pray for people with balance problems. An old woman prayed for Margene. I still wasn't used to this. After all, MS was a gradual deterioration of the myelin sheath insulating the nerves, and I knew from biology classes that nerves didn't regenerate themselves, at least for most species, and for those that did it was a very slow process.

That evening the speaker asked those who had been prayed for earlier in the day to check themselves out to see if their conditions had changed. Margene had slipped away to the back of the sanctuary, and I hadn't paid much attention. Then there was a commotion. I glanced back to see Margene dancing back and forth on the padded chairs that served as pews. When she loudly said that she had been healed, the speaker asked her to come up to the front. I knew as she passed me that she would attempt to pirouette across the stage. And I was sure that she would fall despite her ballet training. She hadn't been able to dance for years.

She *did* attempt to pirouette across, but to my amazement and the amazement of many who knew her, she pirouetted across one way and then back again. Then she shared part of her story, that this was the next installment of healing that she understood God to have promised her several years earlier.

I did not know then what to make of this abrupt change back to a ballet dancer once again. It did not fit with my previous Lutheranism, and it did not fit with my science, and so I searched for a way to explain it. Instead of God healing her, I thought maybe she suffered a spontaneous remission just that day. Whatever that meant, it fits better in my mind, if not in my heart.

After Margene's balance returned (I was initially reluctant to say "miraculously healed," as so many did), another symptom of multiple sclerosis disappeared. She sought prayer from the elders of our church, as suggested in Acts. I watched as something spooky seemed to leave her. By then I had been present more than once when she had been instantaneously healed of some of her symptoms of multiple sclerosis. This was not easy for me to discount. I had seen her start to do things she had not been able to for many years.

Although spontaneous prayer-related healing did not always occur, Margene identified several distinct healing events, about which neither biology nor medicine has explanations, and similar such events are repeatedly reported by others. Coincidences? As an atheist statistician friend of mine once confided to me, "There really are no coincidences." Thus, Scripture offers the only explanation that takes such events seriously, that considers them truly possible.

I began to look for patterns, as I had been doing for many things that I had experienced at this new church. First, although the televangelists liked instantaneous healings that could be demonstrated on stage, many reported healings were not instantaneous. Like some of Margene's, they occurred over hours or days or longer periods of time. Second, they mostly seemed to be related to someone praying, usually directly for a person, sometimes while touching someone but other times at a distance. Indeed, some people reported spontaneous healing even without obvious prayer. Some people reported being healed following prayer, but others were not healed in seemingly the same situations. The only pattern was the lack of pattern, and that raised many issues for me. For example, was healing related to some quality of the person praying or of the person receiving prayer? Were some people just more valuable to God? Was God erratic?

Biology seemed of little help in explaining why or how Margene was healed. The doctor who diagnosed her with MS all those years ago had sent her home saying that there was nothing medicine could do for her. Although we've subsequently met some doctors who allow that spontaneous healing sometimes occurs, others were only vaguely tolerant. When Margene reported her healing to her neurologist in Boston, the doctor begrudgingly allowed that "these things sometimes happen." Remission is the technical term, and one type of MS is termed "relapsing-remitting." A plausible explanation, except that her healing occurred suddenly and coincident with prayer.

Science was even less helpful because the inconsistency made experiments difficult. The first principles needed for deduction and the patterns needed for induction were missing. If the reported healing could be confirmed by medical tests made before and after prayer, then perhaps. But scientists have been reluctant to attempt such comparisons, especially because of the possible dependence of the results on some undefined qualities of the people involved. Not a good candidate for a controlled experimental study, particularly if you'd like to get grant money.

Scripture was seemingly more helpful than science, with many descriptions of spontaneous healing in the Old Testament and even more in the New Testament. Jesus is reported as having healed one or more people in more than 25 separate events, doing so reportedly through prayer, forgiveness, physical touch, and love. Further, many of his disciples also recorded that they saw people healed.

While I had struggled with these issues, with the sudden healing of my anger toward authority I now began to believe that healing could happen spontaneously, healing of both physical and emotional issues. I didn't believe this easily, but I had observed and experienced enough that I could no longer deny that something was happening that science could not explain.[1]

Miracles are contentious because their existence implies that reality, the "external reality" of James Sire, is an open-system rather than closed-system. That is, there is something going on that involves more than cause and effect in the up, down, and sideways of our sensory world. I had long felt discouraged from asking about miracles by both scientists and theologians. But having been healed of some issues with authorities and having seen Margene healed, it was time to ask about the truth.

---

1. Deere, *Surprised by the Voice*; idem, *Surprised by the Power.*

14

Freedom to Change

I REMEMBER BEING CONCERNED about truth-telling on the day the fire marshal forbid the Boiler Room Boys from studying science and philosophy in my grade school's boiler room. There were the obvious lies that my teacher and I told the marshal to minimize his concerns, perhaps protecting the teacher's position. But there was another less obvious lie told that day when the teacher used the fire marshal's restrictions rather than voice his own restrictions, in effect ensuring that we asked no more questions about Karl Marx.

Free to Ask Questions

Asking questions has often been discouraged by people in authority. Sometimes the discouragement involved simple redirection, for example, when my high school guidance counselor suggested a class in Canadian history as a replacement for my request for one in Russian history. Sometimes the discouragement was by humiliation, as when my science teacher discouraged me from asking about Wegner's theory of continental drift.

While being forbidden from asking certain questions isn't the same kind of lie as we had told the fire marshal, it nonetheless felt wrong. However, what was more wrong was my refusal to continue to ask my questions; that was a lie to myself. Also, I didn't speak out later when the school counselor suggested a class in Canadian history though I had asked for one in Russian history. I just pretended like my original question wasn't important. It was. On many occasions, I had failed to pursue questions that I had raised when I felt or feared the displeasure of authority figures.

Historically, Christianity and the Judaism that it arose from held very strong beliefs about the importance of freely asking questions and speaking one's mind. For example, the Passover seder meal was a night of questioning. A twentieth-century rabbi who walked with Martin Luther King Jr. in Selma in 1965 identified the importance of asking questions: "We are closer to God when we are asking questions than when we think we have the answers."[1]

Similarly, the Greek word *parrhesia* was used in the New Testament to denote both speaking the truth and freedom of speaking. Jesus' answer to the high priest, "I have spoken openly to the world" (John 18:19), was written in Greek using this word. It was also used to describe Paul and John speaking boldly, amazing people (Acts 4:13).

Parrhesia as a form of speaking has been contrasted to rhetorical speaking by Michel Foucault (1926–1984).[2] He argued that parrhesia was speaking freely and truthfully without making generalizations, while rhetoric was aimed at manipulation and generalization, whether or not the speaker actually believed what he was saying. Later, Foucault summarized parrhesia as a form of speech in which:

> the speaker uses his freedom and chooses frankness instead of persuasion, truth instead of falsehood or silence, the risk of death instead of life and security, criticism instead of flattery, and moral duty instead of self-interest and moral apathy.[3]

Thus parrhesia requires freedom and truth-telling. My repeated silences in the face of pressure to not ask my questions were lies to me and to others just as much as falsifying the data in my science fair project had been.

As must be clear, I have failed to ask, or to pursue, many important questions that I came upon. I am tempted to list some of them, but not all are as important as they seemed at the time. Parrhesia is useful for finding our own way through the many questions that arise, where we are free to ask and where others are free to answer us as they understand the truth.

---

1. Abraham Joshua Heschel, quoted at https://reformjudaism.org/blog/2017/04/06/how-sacred-art-questioning-can-lead-us-freedom.

2. Foucault, *Discourse and Truth*, at https://foucault.info/parrhesia/foucault.DT1.wordParrhesia.en/.

3. Foucault, *Fearless Speech*, 20.

Freedom to Ask the Big Questions

I had struggled to ask the big questions since childhood, initially without much encouragement either from my church or from scientists. But eventually, I found James Sire's book *The Universe Next Door*, to be invaluable in sorting out my personal answers. Sire began by identifying eight basic questions,[4] many of the big questions my brother Eric had challenged me with but also some he and I had not thought of. Importantly, by the end of his book, Sire had boiled his questions down to only a few.[5] Those questions were the important ones that he believed could guide us in choosing our own worldview.

Having chosen different paths in my search, I was long fascinated with another poem by Robert Frost, "The Road Not Taken."[6] Frost's character laments having to make a choice of roads[7] and doubts that he would return to a road not initially chosen:

> Oh, I kept the first for another day! Yet knowing how way leads on to way, I doubted if I should ever come back.

My "road not taken" had been Christianity, and I had long been happy not having to take that road seriously. But hearing God's voice and experiencing his Holy Spirit and understanding his love gave me the freedom to reexamine my choices. What I've found is that science and Christianity are not mutually exclusive as my experiences with the Boiler Room Boys had left me to believe, and that we can reconsider our earlier choices, that is, we can change our minds, change ourselves.

Pop had long claimed that "People never change, can't change." However, I now saw that he had been wrong—a hard thing to say about your father. I understood that I had changed, but I could not have explained to him how that happened.

So this became my next big question: how is it that people change? This time I did get some help, first by my pastor, Bill Johnson (1951–). In his book *The Supernatural Power of a Transformed Mind*,[8] he focused on being changed by the renewal of our minds in Romans 12:2:

---

4. Sire, *Universe Next Door*, 22–23.

5. Ibid., ch. 11.

6. Frost, "Road Not Taken."

7. Frost explained that the poem was based on a friend who was always be sorry that he hadn't take the other path. Thompson, *Years of Triumph*.

8. Johnson, *Supernatural Power*.

Freedom to Change

> Do not be conformed to this world, but be transformed by the renewal of your mind, that by testing you may discern what is the will of God, what is good and acceptable and perfect.

There he pointed at the "more" just beyond my mind, beyond my autonomous reason. That is part of how people are changed.

I also got help from the cantor at a good friend's recent wedding when he chose the translation by J. B. Phillips (1906–1982) to read 1 Corinthians 13:2:

> If I have the gift of foretelling the future and hold in my mind not only all human knowledge but the very secrets of God . . . but have no love, I amount to nothing at all.[9]

The cantor read the passage it in a manner that allowed me for the first time to really understand the synergism between love and human knowledge. Love gives human knowledge its reason to be and gives us our reason to be.

Together these two verses came together to help me realize that love exists as surely as the physical world exists but that love exists beyond that physical world. It is through recognizing and experiencing the existence of that love that people can change.

---

9. Phillips, *New Testament in Modern English*.

Epilogue: Back to the Boiler Room

I WAS SITTING ON the far end of the couch that evening, watching like a cat, as the young man standing by the fireplace spoke encouraging words to several people. Another man I did not know, sitting on the other end of my couch, said to the group, "I see in my mind's eye rows and rows of flashing lights there," pointing over my head. "They are like computer displays from old movies."

Margene and I had been invited to meet a young man at the home of one of our pastors. I was not very comfortable with him; he seemed a little spooky.

"I know what those lights mean," the young man continued and stepped away from the fireplace, around the coffee table, and I jerked back as he suddenly stood directly in front of me.

"It's coming full circle," he said.[1] "It's coming around to the point where science is starting to catch on that this is all pointing to God. It's not just science; it's really more than that," he continued. "It's thought in general."

I had never met this man, and as far as I knew, he knew nothing about me. I wondered why he had singled me out to talk about science. He couldn't have known that I had been struggling with science and religion for most of my life, and especially in the past several years?

As he began to get into a cadence, I wondered if had chosen me at random and what he was saying was just something on his mind. Perhaps I just happened to be in his way? Or maybe not?

He continued, "God invented our lives, he invented us to work the way we do, and he invented us to create and think and conceptualize and theorize and equate and conclude. He created all those thought processes."

I had not expected that his words would be so close to my heart, to the ideas and feelings that I had been ignoring since God asked me, "Yes or no?" It felt as if he were seeing into my soul, and I became even more afraid.

---

1. Transcribed from a recording made that evening, held by the author.

"You're a thinker, but you're a thinker who has come to the right conclusion. And so many thinkers don't come to the right conclusion, and unfortunately, thinking a lot of times . . . in churches is not encouraged because thinking has been so dangerous to the church for a long time, but, you know, thinkers will catch up."

The young man's words seemed specific to my concern over these last several years, as I had felt myself drifting away from the science that had fascinated me in the boiler room back in grade school. He continued to encourage me of the importance of developing a basis not just for science, but for understanding our thought processes. He then cautioned me that the spiritual and the intellectual must be drawn together and reassured me that my thinking was okay:

> Know that there's nothing inherently wrong with thinking the way you do naturally, but just like any positive trait, it has its potential negatives. God has this symphonic nature of thought and spiritual flow together in you. I think you are going to get a sense of that as you go extending logic to encompass the illogical. As we learn more of God, it's okay for God to be logical. It's like, once we've discovered an area of God and can logically understand that then that's okay because there's an infinite field past that that is completely confusing and mysterious, and we can traverse out into that unexplored territory and eventually stick a flag there, too, and say, "Okay, this is logical now."

The image of exploring territory was okay, but then he said that he had a metaphor for me, one that sounded wrong: "conquering God." He went on to talk about healing, and I kept wondering, "How did he get to healing?" I had been struggling with Margene's physical healing and, more recently, with my own emotional healing of problems with authority. I thought I was done with that. But he continued:

> Where we know exactly why someone gets healed, that's okay because God is not going to run out of tricks. He's infinite, and he's always going to have something new and confusing. That's part of what being in heaven really is. It's looking at those aspects and fully understanding them when you look at them but knowing that there's so much more to understand. I really feel that this is something that not only is growing in you but something you can grow, something you can bring to the rest of the church.

Looking back now, I understand this metaphor of "conquering God," specifically in terms of the words of David (1040–970 BC) in Proverbs 25:2: "It is the glory of God to conceal things, but the glory of kings is to search things out." Conquering God is not challenging his authority but engaging his creation using the capabilities that he has given us. This was also addressed by St. Paul when he writes about our seeking God through his creation (Rom 1:20): "For his invisible attributes, namely, his eternal power and divine nature, have been clearly perceived, ever since the creation of the world, in the things that have been made."

But "conquering God" also reminds me of Jacob wrestling with the angel at Peniel (Gen 32:28), where he won the name Israel: "for you have striven with God and with men, and have prevailed." Here the conquering was much more personal and at this level changes one's heart as well as one's mind.

The young man's prophetic word that evening drew me back toward science, toward the boiler room, toward exploring the world and toward asking how we know about it. I had early experienced the joy of understanding, and then had placed my confidence in my mind, in reasoning through the sensory data that we have the ability to collect. Now my pursuit of God, "conquering God," involved science but also was personal, as Jacob had experienced.

While I don't have my life stretching out before me to pursue answers to life's big questions, I am far better equipped now for this task than I was when I first entered the boiler room. I am freer now to address the important questions, those that challenge the science and religion that I have known. The boiler room has provided and continues to provide a wonderful space for exploring the world and the cosmos that fascinates me.

I have seen others hesitate to enter the boiler room, perhaps listening to their own fire marshals telling them its unsafe. But I've lived my life there and found God isn't afraid of it. In fact, he was right there with me, encouraging me to ask those hard questions, and helping me find some answers. I have written all this with the hope that others, especially those in the ministry, will join me there[2] and learn how to process those same doubts and questions along with learning how to help those in the church do the same.

---

2. www.theboilerroomboys.com.

Glossary

**All definitions adapted from: http://www.merriam-webster.com/
unless noted by an asterisk.**

Abductive logic*: Inference to determine explanations that are consistent with and to eliminate those that are inconsistent with one or more observations

Amino acids: Chemical compounds that have an amino group in the alpha position and that are the chief components of proteins and are synthesized by living cells or are obtained as essential components of the diet

Atheism: A disbelief in the existence of a deity

Axiom: A statement accepted as true as the basis for argument or inference

Big Bang theory: A theory in astronomy that the universe originated billions of years ago in an explosion from a single point of nearly infinite energy density

Born again: Of, relating to, or being a usually Christian person who has made a renewed or confirmed commitment of faith, especially after an intense religious experience

Cambrian: Of, relating to, or being the earliest geologic period of the Paleozoic Era or the corresponding system of rocks marked by fossils of nearly every major invertebrate animal group

Catastrophism: A geological doctrine that changes in the earth's crust have in the past been brought about suddenly by physical forces operating in ways that cannot be observed today

Christianity: The religion derived from Jesus Christ based on the Bible as sacred scripture and professed by Eastern Orthodox, Roman Catholic, and Protestant church bodies

Glossary

Cosmic background radiation: The microwave radiation pervading the universe that exhibits a corresponding blackbody temperature of 2.7 K and that is the principal evidence supporting the Big Bang theory

Cosmological constant: A constant term used in the relativistic equations for gravity to represent a repulsive force that may account in part for the rate of expansion of the universe

Cosmology: A branch of astronomy that deals with the origin, structure, and space-time relationships of the universe

Creationism: A doctrine or theory holding that matter, the various forms of life, and the world were created by God out of nothing and usually in the way described in Genesis

Cretaceous: Of, relating to, or being the last period of the Mesozoic Era characterized by continued dominance of reptiles, emergent dominance of angiosperms, diversification of mammals, and extinction of many types of organisms at the close of the period

Deductive logic: Inference in which the conclusion about particulars follows necessarily from general or universal premises

Deism: A movement or system of thought advocating natural religion, emphasizing morality, and in the eighteenth century denying the interference of the Creator with the laws of the universe

DNA: Any of the various nucleic acids that are usually the molecular basis of heredity, are constructed in a double helix held together by hydrogen bonds between purine and pyrimidine bases which project inward from two chains containing alternate links of deoxyribose and phosphate, and that in eukaryotes are localized chiefly in cell nuclei

Evolution: A process of continuous change from a lower, simpler, or worse state to a higher, more complex, or better state

Exegesis: An explanation or critical interpretation of a text

Existentialism: A chiefly twentieth-century philosophical movement embracing diverse doctrines but centering on analysis of individual existence in an unfathomable universe and the plight of the individual who must assume ultimate responsibility for acts of free will without any certain knowledge of what is right or wrong or good or bad

General relativity: A theory that is based on the two postulates: (1) that the speed of light in a vacuum is constant and independent of the

Glossary

source or observer, and (2) that the mathematical forms of the laws of physics are invariant in all inertial systems, and which leads to the assertion of the equivalence of mass and energy and of change in mass, dimension, and time with increased velocity

Geocentric*: Theory that the earth and other planets rotate around the sun

Geoheliocentric*: Theory that the sun and other planets rotate around the earth

Heliocentric*: Theory that the sun rotates around the earth

Hermeneutics: The study of the methodological principles of interpretation (as of the Bible)

Inductive logic: Inference of a generalized conclusion from particular instances

Jurassic: Of, relating to, or being the period of the Mesozoic Era between the Triassic and the Cretaceous or the corresponding system of rocks marked by the presence of dinosaurs and the first appearance of birds

Magnetometers: An instrument used to detect the presence of a metallic object or to measure the intensity of a magnetic field

Monism: A view that there is only one kind of ultimate substance

Napalm: A thick substance that contains gasoline and that is used in bombs that cause a destructive fire over a wide area

Natural selection: A natural process that results in the survival and reproductive success of individuals or groups best adjusted to their environment and that leads to the perpetuation of genetic qualities best suited to that particular environment

Naturalism: A theory denying that an event or object has a supernatural significance; specifically, the doctrine that scientific laws are adequate to account for all phenomena

Nihilism: A doctrine that denies any objective ground of truth and especially of moral truths

Pantheism: A doctrine that equates God with the forces and laws of the universe

GLOSSARY

Peyote: A hallucinogenic drug containing mescaline that is derived from peyote buttons and used especially in the religious ceremonies of some American Indian peoples

Positivism: A theory that theology and metaphysics are earlier imperfect modes of knowledge and that positive knowledge is based on natural phenomena and their properties and relations as verified by the empirical sciences

Precambrian: Of, relating to, or being the earliest era of geological history or the corresponding system of rocks that is characterized especially by the appearance of single-celled organisms and is equivalent to the Archean and Proterozoic Eons

Quantum mechanics: A theory of matter that is based on the concept of the possession of wave properties by elementary particles, that affords a mathematical interpretation of the structure and interactions of matter on the basis of these properties, and that incorporates within it quantum theory

Quaternary: Of, relating to, or being the geological period from the end of the Tertiary to the present time or the corresponding system of rocks

String theory: A theory in physics that all elementary particles are manifestations of the vibrations of one-dimensional strings

Theism: Belief in the existence of a god or gods; specifically, belief in the existence of one God viewed as the creative source of the human race and the world who transcends yet is immanent in the world

Uniformitarianism: A geological doctrine that processes acting in the same manner as at present and over long spans of time are sufficient to account for all current geological features and all past geological changes

Universe: The whole body of things and phenomena observed or postulated

Worldview (*Weltanschauung*): A comprehensive conception or apprehension of the world, especially from a specific standpoint

Bibliography

Asimov, Isaac. *I, Robot*. New York: Random House, 1952.
Becker, Adam. *What Is Real?: The Unfinished Quest for the Meaning of Quantum Physics*. New York: Basic Books, 2018.
Bhattacharjee, Yudhijit. "Scientist-Politician-Atheist Offers Own Money for Origin of Life Prize." *Science Magazine*, June 27, 2011. http://www.sciencemag.org/news/2011/06/scientist-politician-atheist-offers-own-money-origin-life-prize.
Blackwell, Antoinette. *The Sexes Throughout Nature*. New York: Putnam, 1875.
Blackaby, Henry, Richard Blackaby, and Claude King. *Experiencing God: Knowing and Doing the Will of God*. Nashville: Lifeway, 2007.
Bragg, William H. *The World of Sound*. Adelaide: University of Adelaide Press, 1920. Reprint, London: Andesite, 2015.
Braat, James D. *Abraham Kyper: Modern Calvinist, Christian Democrat*. Grand Rapids: Eerdmans, 2013.
Brown, Nancy M. *The Abacus and the Cross: The Story of the Pope Who Brought the Light of Science to the Dark Age*. New York: Basic Books, 2010.
Chambers, Robert. *Vestiges of the Natural History of Creation*. London: John Churchill, 1844.
Clelland, C. E., and C. F. Chyba. "Defining Life." *Origins of Life and Evolution of the Biospheres* 32, no. 4 (2002) 387–93.
Cohen, Norman. *Noah's Flood*. New Haven, CT: Yale University Press, 1999.
Collins, C. John. *Science and Faith: Friends or Foes?* Wheaton, IL: Crossway, 2003.
Conway Morris, Simon. "Evolution and Convergence." In *The Deep Structure of Biology*, edited by Simon Conway Morris, 46–67. West Conshohocken, PA: Templeton Foundation, 2008.
Coon, Carleton. *The Origin of Races*. New York: Knopf, 1962.
Craig, William Lane. *Does God Exist?* Pine Mountain, GA: Impact 360, 2014.
Crane, Stephen, *Prose and Poetry*. 1899. Reprint, New York: Library of America, 1984.
———. *Reasonable Faith: Christian Truth and Apologetics*. Wheaton, IL: Crossway,1994.
Dalrymple, G. Brent. *Ancient Earth, Ancient Skies*. Stanford: Stanford University Press, 2004.
Darwin, Charles. *Darwin Correspondence Project*. http://www.darwinproject.ac.uk/letter.
———. *The Descent of Man, and Selection in Relation to Sex*. London: John Murray, 1871.
———. *Origin of Species*. London: John Murray, 1859. http://darwin-online.org.uk/content/frameset?itemID=F373&viewtype=text&pageseq=1.

BIBLIOGRAPHY

Davenport, C. B., and M. T. Scudder. "Naval Officers, Their Heredity and Development." Washington, DC: Carnegie Institution, 1919. https://archive.org/details/navalofficersthoodavegoog.

Deere, Jack. *Surprised by the Power of the Spirit*. Grand Rapids: Zondervan, 1993.

———. *Surprised by the Voice of God*. Grand Rapids: Zondervan, 1996.

DeMar, Gary. *The Philosophy of Meaninglessness*. 2008. http://americanvision.org/1071/philosophy-of-meaninglessness/.

Dennett, Daniel. *Darwin's Dangerous Idea*. New York: Simon and Schuster, 1995.

Desmond, A., and J. Moore. *Darwin's Sacred Cause: Race, Slavery and the Quest for Human Origins*. Chicago: University of Chicago Press, 2009.

"Determining the Age of the Universe." *Scientific American*, April 2013.

Dyson, Freeman. *Disturbing the Universe*. New York: Basic Books, 1979.

Eccles, John. *Evolution of the Brain: Creation of the Self*. Abingdon, UK: Routledge, 1989.

Eisley, Loren. *Darwin's Century*. New York: Anchor Books, 1952.

Farber, Stephen. "U.S. Scientists' Role in the Eugenics Movement (1907–1939): A contemporary Biologist's Perspective." *Zebrafish* 5, no. 4 (2008) 243–5.

Fee, Gordon, and Douglas Stuart. *How to Read the Bible for All Its Worth*. Grand Rapids: Zondervan, 2003.

Foucault, Michel. *Discourse and Truth: The Problematization of Parrhesia*. 6 lectures at the University of California at Berkeley, October–November 1983. https://foucault.info/parrhesia/.

———. *Fearless Speech*. Los Angeles: Semiotext(e), 2001.

Frost, David. *Frost/Nixon: Behind the Scenes of the Nixon Interviews*. New York: Harper Perennial, 2007.

Frost, Robert. "The Lockless Door." 1920. www.poetry-archive.com/f/the_lockless_door.html.

———. "The Road Not Taken." 1920. www.poetry-archive.com/f/the_road_not_taken.html.

Fuller, Buckminster. *Nine Chains to the Moon*. Philadelphia: Lippincott, 1938.

Gould, Stephen J. "D'Arcy Thompson and the Science of Form." In *Topics in the Philosophy of Biology*, edited by Marjorie Greene and Everett Mendelson. Boston Studies in the Philosophy of Science 27. Berlin: Springer, 1976.

Haarsma, Loren, and Terry Gray. "Complexity, Self-Organization and Design." In *Perspectives on an Evolving Creation*, edited by Keith B. Miller 288–311. Grand Rapids: Eerdmans, 2003.

Halafta, Jose ben. *Seder Olam Rabbah*. Translated by Albert Benhamou. http://seder-olam.info/.

Hall, Marshall. *The Earth Is Not Moving*. Athens, GA: Fair Education Foundation, 1991.

Hare, Augustus W., and Julius C. Hare. *Guesses at Truth*. Boston: Ticknor and Fields, 1827. https://archive.org/details/guessesattruthb03haregoog.

Harris, Sam. *The End of Faith: Religion, Terror, and the Future of Reason*. New York: Norton, 2004.

———. *Free Will*. New York: Simon and Schuster, 2012.

Hitchens, Christopher. *The Missionary Position: Mother Teresa in Theory and Practice*. New York: Twelve, 1995.

Hummel, Charles. *The Galileo Connection*. Westmont, IL: InterVarsity, 1986.

Hutton, James. *Theory of the Earth*. Edinburg: Royal Society, 1788.

Bibliography

Huxley, Aldous. *Ends and Means: An Enquiry into the Nature of Ideals and into the Methods Employed for Their Realization.* London: Chatto and Windus, 1937, 1951.

Huxley, Julian. *Evolution: The Modern Synthesis.* Crows Nest, Australia: Allen and Unwin, 1942.

Huxley, Thomas H. *Evidence as to Man's Place in Nature.* London: Williams and Norgate, 1863.

Iliffe, Robert. *Priest of Nature: The Religious Worlds of Isaac Newton.* Oxford: Oxford University Press, 2017.

Isaacson, Walter. *Einstein: His Life and Universe.* New York: Simon and Schuster, 2008.

Johnson, Bill. *The Supernatural Power of the Transformed Mind: Access to a Life of Miracles.* Shippensburg, PA: Destiny Image, 2005.

Josephus, Flavius. *The Antiquities of the Jews.* AD 93. Translated by William Whiston. https://www.gutenberg.org/files/2848/2848-h/2848-h.htm.

Kühl, Stefan. *The Nazi Connection: Eugenics, American Racism, and German National Socialism.* Oxford: Oxford University Press, 1994.

Kipling, Rudyard. *Just So Stories.* Basingstoke, UK: Macmillan, 1902.

Lamarck, Jean Baptiste. *Theory of Inheritance of Acquired Characteristics.* 1801.

Lankester, E. Ray. *Degeneration: A Chapter in Darwinism.* Basingstoke, UK: MacMillan, 1880.

Lewis, C. S. *Surprised by Joy.* San Diego, CA: Harcourt Brace, 1955.

———. *The Collected Letters of C. S. Lewis.* Edited by Walter Hooper. 3 vols. New York: Harper Collins, 2007.

Locke, John. *Essay Concerning Human Understanding.* 1690. Reprint, edited by Jonathan Bennett, 2004. https://www.earlymoderntexts.com/assets/pdfs/locke1690book4.pdf.

Lombardo, Paul A. *A Century of Eugenics in America: From the Indiana Experiment to the Human Genome Era.* Bloomington: Indiana University Press, 2011.

Lyell, Charles. *Principles of Geology: Being an Attempt to Explain the Former Changes of the Earth's Surface, by Reference to Causes Now in Operation.* New York: Appleton, 1830. https://www.gutenberg.org/files/33224/33224-h/33224-h.htm.

McDougall, I., F. H. Brown, and J. G. Fleagle. "Stratigraphic Placement and Age of Modern Humans from Kibish Ethiopia." *Nature* 433 (2005) 733–36.

Macdougall, J. D. *Why Geology Matters: Decoding the Past and Anticipating the Future.* Berkley: University of California Press, 2011.

Malthus, Thomas R. *An Essay in the Principle of Population.* London: J. Johnson, 1798.

Mauer, Raymond J. *Duck and Cover.* 1952. https://www.youtube.com/watch?v=lcl8uRn6uZQ.

Miller, Stanley L. "A Production of Amino Acids Under Possible Primitive Earth Conditions." *Science* 117, no. 3046 (1953) 528–29.

Mora, Camilo, Derek P. Tittensor, Sina Adl, Alastair G. B. Simpson, and Boris Worm. "How Many Species Are There on Earth and in the Ocean?" *PLoS Biology* 9, no. 8 (2011). https://doi.org/10.1371/journal.pbio.1001127.

Morris, Henry. *Science and the Bible.* Chicago: Moody, 1986.

Nagle, Thomas. "A Philosopher Defends Religion." *The New York Review of Books*, September 27, 2012.

———. *Mind and Cosmos: Why the Materialist Neo-Darwinian Conception of Nature Is Almost Certainly False.* Oxford: Oxford University Press, 2012.

BIBLIOGRAPHY

Orestes, Naomi. *Plate Tectonic: An Insider's History of the Modern Theory of the Earth.* Boulder, CO: Westview, 2003.

Orestes, Naomi, and John Krige. *Science and Technology in the Global Cold War.* Boston: MIT Press, 2014.

Pearcey, Nancy, and Charles Thaxton. *The Soul of Science.* Wheaton, IL: Crossway, 1994.

Pereto, Juli, Jeffrey L. Bada, and Antiono Laxcano. "Charles Darwin and the Origin of Life." *Origin of Life and Evolution of the Biosphere* 39, no. 5 (2009) 395–406.

Peterson, Eugene. *The Message: A Contemporary Language Translation of the Bible.* Colorado Springs, CO: NavPress, 2002.

Phillips, J. B., translator. *The New Testament in Modern English.* New York: Simon and Schuster, 1958.

Pigliucci, Massimo, and Gerd Muller. *Evolution: The Extend Synthesis.* Cambridge, MA: MIT Press, 2010.

Plantinga, Alvin. "Why Darwinist Materialism Is Wrong." *The New Republic*, November 16, 2012.

———. *Where the Conflict Really Lies: Science, Religion and Naturalism.* Oxford: Oxford University Press, 2012.

Quammen, David. *Song of the Dodo: Island Biogeography in an Age of Extinctions.* New York: Simon and Schuster, 1996.

Ross, Hugh. *Navigating Genesis: A Scientist's Journey through Genesis 1–11.* Covina, CA: Reasons to Believe, 2014.

Schopf, J. William. *Cradle of Life: The Discovery of the Earth's Earliest Fossils.* Princeton, NJ: Princeton University Press, 1999.

Serre, David, and Savante Paabo. "Evidence for Gradients of Human Genetic Diversity within and Among Continents. *Genome Research* 14 (2004) 1679–85.

Sire, James W. *Apologetics Beyond Reason: Why Seeing Really Is Believing.* Westmont, IL: InterVarsity, 2014.

———. *The Universe Next Door.* 5th ed. Westmont, IL: InterVarsity, 2009.

Smith, Tim D. *Scaling Fisheries: The Science of Measuring the Effects of Fishing.* Cambridge: Cambridge University Press, 1994.

Sofair, A. N., and L. C. Kaldjian. "Eugenic Sterilization and a Qualified Nazi Analogy: The United States and Germany, 1930–1945." *Annals of Internal Medicine* 132, no. 4 (2000) 312–19.

Spencer, Herbert. *Principles of Biology.* London: Williams and Norgate, 1864.

Statistic Brain Research Institute. "Bible Statistics." http://www.statisticbrain.com.

Stenger, Victor J. *God: The Failed Hypotheses.* New York: Prometheus, 2007.

Sugitani, Kenichiro, et al. "Biogenicity of Morphologically Diverse Carbonaceous Microstructures from the ca. 3400 Ma Strelley Pool Formation, in the Pilbara Craton, Western Australia," *Astrobiology* 10, no. 9 (2010) 899–920.

Thompson, D'Arcy. *Growth and Form.* Cambridge: Cambridge University Press, 1917, 1942.

Thompson, Francis. "The Hound of Heaven." 1890. https://www.ewtn.com/library/HUMANITY/HNDHVN.HTM.

Thompson, Lawrance. *Robert Frost: The Years of Triumph.* Minneapolis: University of Minnesota Press, 1971.

Tirard, S., M. Morange, and A. Lazcano. "The Definition of Life: A Brief History of an Elusive Scientific Endeavor." *Astrobiology* 10 (2010) 1003–9.

Bibliography

Wallace, Alfred. "Geological Climates and the Origin of Species." *Quarterly Review* 126 (1869) 359–94.

Wertheim, Margaret. "God Is also a Cosmologist." *New York Times*, June 8, 1997.

White, Andrew D. *A History of the Warfare of Science with Theology in Christendom*. New York: Macmillan, 1896.

Wilczek, Frank. *A Beautiful Question: Finding Nature's Deep Design*. New York: Penguin, 2015.

Woit, Peter. *Not Even Wrong: The Failure of String Theory and the Search for Unity in Physical Law*. New York: Basic Books, 2006.

Young, Davis A. *The Biblical Flood: A Case Study of the Church's Response to Extrabiblical Evidence*. Grand Rapids: Erdmans, 1995.

Wittgenstein, Ludwig. *Tractatus Logico-Philosophicus*. Translated by C. K. Ogden. New York: Harcourt Brace, 1922. https://www.gutenberg.org/files/5740/5740-pdf.pdf.

Subject Index

abstract reasoning, 140–43
acquired characteristics, 115–17
Acts, 70, 95, 110, 147; *in Bible*
Adam, 83, 139–42; *in Bible*
Africa, 17, 77–78, 108, 134–35
African Americans, 133–35
agnostic, 28, 94
Al-Khwarizmi, Muhammad ibn Musa, 6
Alexander the Great, 83
algae, 23
algebra, 6, 15
altruism, 142
American, 3–4, 7, 20, 30, 54, 125, 132
American Breeders Association, 137
amino acids, 3, 21, 102, 105
amoebas, 108
Ananias and Sapphira, 70; *in Bible*
Anning, Mary, 110–11
anthropological, 133, 135, 138
ants, 108, 141
Aramaic, 35
archaeological, 35, 135
Arctic Ocean, 50–51
Aristotle, 4, 74, 104
Asimov, Isaac, 36
Aspidella terranovica, 111
asteroid, 111
atheist, 28, 33–34, 38, 93–95, 99–100, 105, 125, 128, 139
Atlas Mountains, 108
atoms, 33, 55–56
authority, 14, 68, 70, 83, 90, 94, 144

B-29 Superfortress, 2–3, 7, 144
Babel, Tower of, 134; *in Bible*

baby boom, 19–20, 23
Babylonians, 73
baptism, 46–47, 97–98
beagles, 109
Beatles (music group), 32
Becker, Adam, 60
Becquerel, A. Henri, 85
beetles, 40, 109
Bell, Alexander Graham, 136
beryllium, 86
Bessel, Fredrich, 74
Bible, 12, 26–27, 35–36, 57, 73–75, 81–84, 97, 107, 128–30
Big Bang, 55–58, 96
big questions, 1, 4, 11–12, 27, 29
biomathematics, 43, 47
birds, 101, 107, 109, 111, 118
Blackaby, Henry, 70
Blackwell, Antoinette, 138–39
Blake, William, 25
blending inheritance, 22, 115–17
blue whales, 23, 43–44
Bohr, Nils, 59
Boiler Room (Boys), 1–6, 12, 15, 24, 30, 53, 63, 77, 99, 102, 145–46, 150, 152, 154–56
bombs
  air raid drills, 3–5
  Fat Man, 59
  firebombs, 2–4, 41, 78
  Little Boy, 59
  plutonium, 3
  shelters, 3
  Trinity explosion, 59
bombs (continued)
  uranium, 3

Subject Index

war ending, 3
born again, 67–68
Born, Max, 59
Bragg, William Lawrence, 124
Brahe, Tycho, 74
Bretz, J. Harlan, 82
Buddhism, 32, 97
Burnet, Thomas, 80
Burton, Richard, 88

Cambrian Explosion, 111
Camp Creek Valley, Oregon, 1–2, 21, 40, 73, 91, 122, 132, 135
Canyon Diablo meteorite, 86
Cape Cod, Massachusetts, 121, 144
Carnegie Foundation, 136–37
Chambers, Robert, 113
charismatic, 67–68, 70
chemistry, 2, 16, 102
Chicxulub, Yucatan Penninsula, 111
Chihuahua dogs, 112
chimpanzees, 132
China, 16, 20, 58, 73, 134
Christian theists, 34, 39, 125
chokmah (wisdom), 57; *in Bible*
Churchill, Winston, 136
Clark River, 82
cod fish, 23, 43, 68, 71, 121, 144
cognitive faculties, 34, 130, 132, 138–40
Collins, C. John, 14, 16, 18, 36, 83–84, 116, 130
Confucianism, 97
Connecticut State, 91, 94, 121–22, 124
conquering God, metaphor, 155–56
constellations
    Bear, 73, 118
    observed by Chinese, 73
    Orion, 74
    Pleiades, 74
continental drift, 77, 87
continents, 17–18, 73, 76–79, 87, 134
Coon, Carleton, 133
Copernicus, 74–75
cosmic microwave background radiation, 54–55
cosmology, 17, 54–55, 60, 118
counselor, 8, 64–66

crabs, 41
Crane, Stephen, 29, 33
creation of the earth
    analogical-day view, 84
    day-age view, 84
    literary framework view, 84
    normal-day, 84
    old earth creationism, 130
    young earth creationism, 130, 135
crocus, 23
Curie, Marie, 85
Curie, Pierre, 85

Daniel, 74, 126, 129; *in Bible*
Darwin, Charles, 21–22, 24–25, 81, 103–5, 108–9, 112–18, 127, 133–36, 138–42
Davenport, Charles B., 136
Dawkins, Richard, 125
Dennett, Daniel, 126
Descartes, Rene, 36
determinism, 36–37, 127, 135, 138
Diderot, Denis, 36
Dietz, Robert, 79
Dilthey, Wilhelm, 28
dinosaurs, 110
DNA, 103, 105, 107, 118–19, 131–32, 134–35
dog breeders, 112
dog vomit fungus, 10
dolphins, 61–65, 68, 71
Dorset coast of United Kingdom, 110
Duck and Cover, 3, 8
ducks, 123
duckweed, 108
Dyson, Freeman, 57

earth, 1, 4, 16–17, 54, 57, 73–74, 80, 101, 104–7, 115, 119, 124, 134
    age of, 83–87, 125, 128
    early conditions on, 3, 102, 105
    flat, 74, 129
    floods and earthquakes, 76, 80–81
    formation of, 75–77
    layers in crust, 76, 110–11, 157
    magnetic polarity, 78–79
    moving, 74–75, 128–29, 159
    shadow on Venus, 74

Subject Index

unchanging, 17, 73, 87
ecardia, 111
Einstein, Albert, 7
Eisenhower, President Dwight, 7–8, 10, 12
Eisley, Loren, 141
elements, 2–3, 5–6, 15–16, 85–86, 102, 102, 108
elohiym (God), 57–58, 119, 139; *in Bible*
empathy, 142
Eniwetok Atoll, Marshall Islands, 40–41, 64
Episcopal Church, 68, 70, 92
ethics, 30, 34, 142
Euclid, 4, 6–7, 15–16, 23
eugenics, 136–38
Euphrates River, 134
Eve, 140–42; *in Bible*
evil, 66, 97, 136
evolution, 10, 21, 24–25, 104, 116–19, 125–27, 129, 132–36
   degenerative, 135
   human evolution, 133
   macroevolution, 118
   microevolution, 118, 135
   progressive, 133, 137
exegesis, 35
experiments, 3, 8–11, 13–16, 22, 24–25, 48, 60, 104–5, 116, 136
external reality, 30, 33

faith, 14, 28, 31–32, 36, 53–54, 57, 83–84, 100, 104, 116, 119, 125, 130
families, 42, 48, 66, 107, 135–36, 142
Fibonacci sequence, 23
finches, 109, 118
fire marshal, 5, 8, 10, 12, 63
fishermen, 61–65, 68–69, 71
fishery biology, 69
FitzRoy, Robert, 108
Fletcher's Ice Island, 50–51, 61
floods, 17, 76, 79–82, 87, 110
forced sterilizations, 137
forces, 13, 24, 96, 105, 107, 132, 134
   catastrophic, 76, 82
   electromagnetic, 56
   gravitational, 7, 33, 54, 56, 85
   mechanical, 99, 102
   nuclear, 4, 41, 55–56, 58–59, 78, 85
   strong nuclear, 56
   vital, 102
   weak nuclear, 56
fossils, 108–11, 114–15, 117
   in layers, 111
   in meteorites, 104
   human, 134
   marine fossils, 17, 81
   microfossils, 102, 111
   single-celled fossils, 119
Foucault, Michel, 151
freedom, 92, 98, 123, 130, 142, 158
   Biblical basis for, 100
   ethics and, 142
   free will, 99
   to choose, 100, 120, 139, 141–42
   to speak, 98, 151–52
Frost, Robert, 93
funambulist (rope walker), 31

Galapagos Islands, 23, 109, 118
galaxies, 17, 54–56, 109, 115
Galileo, 7, 74
Galton, Francis, 115
garden of Eden, 134, 139; *in Bible*
geese, 123
gemmules, 115–16
Genesis, 17, 36, 53, 57–58, 73, 75, 80, 83–84, 101, 106, 109–10, 129–30, 133–34, 139–40; *in Bible*
genetic coding mechanism, 103, 119
geologists, 17, 75 79, 81 82, 85, 87, 111
geological periods, 77, 111
geometry, 6, 15–16
germ cells, 116–17
Germany, 4, 68
giant tortoises, 23
goats, 109, 111–12
God
   all religions point toward, 26
   anger of, 80

God (continued)

169

celebrated through astronomy and
    cosmology, 7
characteristics of, 39, 57, 153,
    155–56
contract with, 91–92, 100, 120
creative activities of, 1, 4, 22, 34,
    57–58, 70, 73, 76, 83, 101, 103–4,
    106, 129, 133, 139, 154, 158
creator, 60, 125
encounters with, 7, 26, 32, 35, 67,
    71, 88–89, 91–93, 95–96, 98, 100,
    123–24, 146, 152
existence of, 94–97, 125–26, 159–60
image of, 37, 96, 140
mind at large, 26, 97
mysteries of, 26, 147, 155
names of, 139
personal relationship with, 66, 100,
    120, 122, 139, 142, 152
promises of, 80, 87
source of special and internal revelation, 34
spirit of, 57, 101, 106, 119, 140
Trinity, 93, 119
uncertainties about, 73, 88
value for people, 34, 51, 97, 100, 148
wisdom of, 57
Gravelet, Jean Francois, 31
Gray, Terry, 105
Great Dane dogs, 112
Greek(s), 6, 35, 53, 73, 98
Green, William Henry, 83
growth, 8–10, 15, 19, 23, 25, 42–43, 62,
    112–13
growth hormone, 8–9, 15

Haarsma, Loren, 105
Hahn, Otto, 104
Halafta, Yose ben, 83
Haldane, J.B.S.
Hall, Marshall, 128–29
Hannah, 74; *in Bible*
Harris, Sam, 99, 125
healing, 68, 70–71, 88, 94
heart, 11, 26, 30, 40–41, 49, 131
Hebrew, 30, 35, 53, 57, 73, 76, 83, 129,
    139
Heisenberg, Werner, 59

heredity, 103, 115–17
hermeneutics, 35
Hess, Harry, 79
Hiroshima, Japan, 3, 6–7, 59, 84
history, 8, 20, 30, 34–35, 69, 75–76, 85,
    104, 109, 113, 119, 124–25, 127,
    129–30, 133
Hitchens, Christopher, 126
HMS Beagle, 108
Holmes, Arthur, 85
Homo sapiens, 132, 134–35
honey bees, 24
Hooke, Robert, 81
house churches, 70
Hoyle, Fred, 3
Hubble, Edwin, 17
human
    evolution, 133
    fossils, 134
    genome, 106, 130, 138
    knowledge, 140, 153
    population, 23
    race, 138
    spirit, 11, 91, 102, 119, 140, 155
husbandry, 109, 111–12, 136
Hutton, James, 76
Huxley, Aldous, 25–28, 30, 38–39, 97
Huxley, Julian, 21–22, 25, 117–18, 137
Huxley, Thomas, 25, 115
hypotheses, 8, 13, 105

ice dams, 82
Iceland, 78, 108
Indonesia, 109
intelligence
    artificial, 37
    extraterrestrial, 106
    higher, 138
Inyan Kara Mountain, 32
Isaiah, 55, 129; *in Bible*
Israel, 134

Japanese, 1, 4, 28, 30, 41, 97, 131
Jeremiah, 128, 142; *in Bible*
Jesus, 1, 4, 12, 26–30, 35, 47, 53, 81,
    91, 93, 97–98, 100, 121, 129, 142;
    in Bible
Jews, 4, 27, 30, 97

Subject Index

John the Baptist, 97–98; *in Bible*
John, St., 53–54, 57, 98, 142, 151; *in Bible*
Johnson, Bill, 152
Joram, 83; *in Bible*
Jordan River, 98
Josephus, Flavius, 26–28, 47, 97–98
Journey of the Dead Man Desert, New Mexico State, 59

Kalam argument, 57–58, 96
Kant, Immanuel, 28
Kepler, Johannes, 7, 75, 83
Keynes, John Maynard, 136
Khrushchev, Premier Nikita, 7, 12
King David, 30; *in Bible*
King Nebuchadnezzar, 74; *in Bible*
King Solomon, 57, 81, 129; *in Bible*
King, Martin Luther, Jr., 42, 151
Kipling, Rudyard, 117
Kline, Meredith, 84
knowledge, 44, 125
   human, 140, 153
   objective, 28, 33
   source of, 28, 30, 33–34, 96, 99, 140
   system of, 13–14
Korea, 2
Kuhn, Thomas, 54
Kuyper, Abraham, 28

Laban, 109; *in Bible*
Lake Missoula, Montana State, 82
Lamarck, Jean Baptiste, 115
Lankester, E. Ray, 135
lava, 78–79, 108
layil (night), 75; *in Bible*
lead, 85–86
leukemia, 47–48
Lewis, C. S., 35–36, 88–89, 92–93, 95, 99, 121, 142
life, 3, 29, 38–39, 46, 50, 56, 58–59, 87, 96–97, 100, 111, 132, 135, 142–43, 145, 151, 154, 156
   chemistry of, 135
   creation of, 1, 3–4, 11, 21, 105–7, 119
   definition of, 22, 24–25, 101–2, 119–20
   diversity of, 22, 24, 107–11, 114–15, 117–19
   meaning of, 37–38, 49, 100, 127
   origin of, 1, 21–22, 102–5, 112, 125–26
   purpose of, 34, 36, 38, 132
   reproduction of, 103, 118
   requirements for, 103, 108
   spiritual aspects of, 60, 80, 94, 100, 102, 119, 139–40
lizards, 10–11, 25
logic
   "a matter of course," 17, 37, 54, 79, 114, 118
   abductive, 16–17, 37, 54
   deductive, 6–7
   inductive, 7, 16
logos (word), 57; *in Bible*
Lombok Strait, 109
Lord Kelvin, 85
love, 4, 26, 98, 137, 143, 148, 152–3
Luke, St., 81, 97; *in Bible*
Luther, Martin, 42, 68
Lutheran Church, 20, 46–47, 67
Lyell, Charles, 76

magnetite, 78–79
Malay Archipelago, 109
Malta, 110
Malthus, Thomas, 23–24, 112–13
mammals, 111, 132
Mantell, Mary Ann and Gideon, 110–11
Marine Mammal Protection Act, 61, 71
Mark, St., 38, 97; *in Bible*
Marx, Karl, 4–5, 8
materialism, 4, 60, 139
mathematical model, 33, 55–56, 58, 74, 97
mathematical singularity, 55, 58
mathematics, 6–7, 15, 21–24, 27, 42–43, 47–48, 61, 140–41
Matthew, St., 97; *in Bible*
McKenzie River, 1, 73, 79, 92
measurement, 9, 23, 59, 78–79, 87, 133, 141
Mediterranean Sea, 108
Mendel, Gregor, 22, 115

Subject Index

Mendeleev, Dmitri Ivanovich, 2
mescaline, 25
meteorites, 86–87, 104
mice, 104, 132
Michelangelo, 140
Mid-Atlantic Ridge, 78–79, 108, 111
military draft, 19–20, 42, 44
Miller, Hugh, 81
Miller, Stanley, 3, 105
miyn (type), 107, 111; *in Bible*
Monopoly (board game), 89, 94
Montana State, 82
moral premise, 96
morality, 39
Morgan, Thomas, 136
Morris, Henry, 84, 128
Moses, 36, 53–55, 57, 73, 75–76, 80,
    83–84, 86, 97, 101–4, 106–7,
    109–11, 129, 134, 139–42; *in Bible*
Mount Everest, 32
Mount Shasta, 32
MS, see multiple sclerosis
Muller, Gerd, 118
multi-worlds, 60
multidisciplinary, 42–43
multiple sclerosis, 65–68, 70, 89, 94
Murray, Alexander, 111
Muslim philosophers, 57
myelin sheath, 66, 89

Nagasaki, Japan, 3, 59
Nagel, Thomas, 139
National Academy of Sciences, 13
Native Americans, 132
natural history, 95, 113
natural selection, 21–22, 113–18, 126,
    138–39
naturalists, 33–34, 38, 125
Nazis, 1, 4, 97, 137
Neo-Darwinism, 139–40
nephesh (breath), 107; *in Bible*
New Mexico State, 3, 59, 121
Newfoundland, 111
Newton, Isaac, 7, 83, 119
Niagara Falls, 31
nihilism, 4, 49
Nile River, 76, 104
Noah's flood, 80–82, 110; *in Bible*

Nobel Prize, 54, 59–60, 94, 124
North America, 17, 68, 71, 77–79, 82,
    111, 134–35
North Atlantic Ocean, 71, 78–79
North Pole, 50–51, 61, 94
North Vietnamese sailors, 19–20

octopus, 114, 118
Oklahoma State, 133, 135
Ortelius, Abraham, 77
Oslo, Norway, 142

Pacific Lutheran University, 20
Pacific Ocean, 20, 41, 69, 82, 122
Pangaea, 78
pangenesis, 115, 117
panspermia, 104–5
Pardee, Joseph T., 82
parrhesia (speaking truth and/or free-
    dom),151; *in Bible*
Patience, card game (or Solitaire), 2, 89
Patterson, Clair, 86
Paul, St., 95, 100, 110; *in Bible*
Pauli, Wolfgang, 59
pea plants, 22, 116
Pearcey, Nancy, 141
Pearson, Karl, 22
Peirce, Charles Sanders, 16
Penzias, Arno, 54
periodic chart of the elements, 2, 5–6
Perry, John, 85
Philippines, 1–2
Phillips, J.B., 153
philosophy
    biological truths from, 24
    charts of, 4–6, 12, 27
    father of modern, 36
    Huxley's perennial, 25–26, 30
    Marx's, 5, 8
    modern system of, 54–55
    names of types of, 4
    of nature, 141
    of social Darwinism, 136
    Spencer's synthetic, 24
phrenology, 133
Pigliucci, Massimo, 118
Pimelia (beetle) 109
placebo, 48

Subject Index

Planck, Max, 54
planets, 7, 74, 87, 106, 112
   Mars, 106
   Pluto, 33
   Saturn, 74
   Venus, 74
Plantinga, Alvin, 126–27, 139–41
plutonium, 3, 8, 59
pneuma (spirit), 98; *in Bible*
Poland, 4
"Pop" (author's father), 1–2, 11, 41–42, 67, 79, 89, 95, 131, 133, 135
population, 10, 21, 23, 43–44, 48, 62, 71, 112–13, 136–37, 142
presuppositions, 30–31, 38, 53–54
Price, George, 81
prime reality, 30, 33
Priscilla and Aquila, 70; *in Bible*
proteins, 102–3, 105, 119
Ptolemy, Claudius, 74

quantum physics, 56, 59–60

rabbits, 11, 23–25, 92–93, 101, 116
race
   Caucasoid, 133
   Congoid, 133
   human, 138
   riots, 133
   theory of, 133, 138
radiation, 3, 54–56, 84–85
radioactive isotopic decay, 85–87
radiometric time clock, 78
radium, 85
raqiya (firmament), 74; *in Bible*
Reagan, President Ronald, 65, 67
reality, 7, 11, 14–15, 26, 30, 33, 60–61, 68, 89, 99, 106, 110
Redi, Francesco, 104
redwood trees, 108
religions, 7, 32, 92, 97, 125, 139
   ambiguity in, 87, 120
   answers to big questions, 12, 56, 146, 156
   distinctions among different, 39, 94
   dogmatism in, 124, 128

   relationship to science, 17–18, 29, 35, 72, 81, 101, 124–27, 130, 141, 154
   scripture, 26, 35, 53, 57, 74–76, 81, 84, 95, 119, 139, 147–48
   touchstones and assumptions, 53, 74–75, 84, 125, 138
remission of disease, 89
renewal of our minds, 152
resurrection, 27, 35
right whales, 71
robots, 36–37, 99
Ross, Hugh, 83
Russia, 2–4, 7–8, 78
Rutherford, Ernest, 85
ruwach (spirit), 57, 119; *in Bible*

Sacramento River, 122–23
Samuel, 74; *in Bible*
San Diego, California State, 61, 63–65, 67–68
scablands, Washington State, 82
Schopf, J. William, 102
science, 2–3, 72, 87–88, 120, 128–30, 146–47, 154–56
   ambiguity in, 54, 87, 89
   answers to big questions, 4, 11–12, 15, 60, 94, 98, 148, 150
   definition of, 13–14, 16, 138, 141
   fiction, 3, 36, 99
   honesty in, 9–10, 30, 65
   mistakes in, 54, 133, 136
   of spiritual phenomena, 28
   relationship to religion, 17–18, 29, 35, 53, 92, 101, 106, 124–28, 130, 138, 140–41, 148, 152
   science fair project, 8–10, 12–14, 24–25, 48, 63, 94, 151
   scientific method, 8, 10, 13
   teachers, 1, 3, 10, 13, 54, 71, 124, 146, 150
   touchstones and assumptions, 6, 12–13, 15–18, 53–55, 60, 75, 94, 106, 125, 138
   scripture, 35, 53, 56–57, 74–76, 81, 84, 95, 119, 139
seals, 44, 71
self-consciousness, 141–42

Subject Index

sensus divinitatis, 140
Seventh Day Adventist Church, 81
sharks, 110
sheep, 109, 111–12
Sire, James, 29–30, 125
Sistine Chapel, 140
skunks, 11
slime mold, 10, 24
Smith, Huston, 97
Smith, Margene (author's spouse), 20–21, 41–42, 45–47, 49, 61, 63–72, 88–92, 94, 121–23
snakes, 110
solar system, 33, 73–75, 87
    eclipses, 73
    geometry
    few clues from Bible, 75
solar system (continued)
    mathematical models of, 74
    meteorites, 86
    other solar systems, 106, 109
Solitaire, card game (or Patience), 2, 89
South America, 17, 77–78, 135
species
    crossbreeding, 109, 111, 115, 117
    degeneration of, 136
    differences among, 107, 111, 116–18
    differences within, 109, 112, 134–35
    spatial distribution of, 77, 108
    extermination, 71
    genera, 107
    geographic barriers, 109
    in fossil record, 110–11, 114
    miyn (type), 107; *in Bible*
    number of, 107, 163
    origin of, 22, 24, 81, 112, 114, 117
    progressive change, 113
    selective breeding, 109, 136
Spencer, Herbert, 24, 102, 115, 119
spirit
    God's, 57, 91–94, 97–98, 101, 106, 119, 140, 145, 152
    human, 11, 91, 102, 119, 140, 155
    spiritual phenomena, 28, 32, 34, 38, 60, 95, 100, 155
sprout, 101, 107
spy plane (U-2), 7–8, 10, 12
stars

61 Cygni, 75
Alpha Centauri, 16
    guest star, 16–17
    named by Chinese, 73
    Pleiades, 74
Stenger, Victor, 38
Steno, Nicolas, 110
Sunday school, 1, 4, 12, 28–29, 73, 90, 96, 98
Surtsey Island, 108
survival of the fittest, 113–14, 135–36

teleological premise, 96
telescopes, 16, 106
Ten Commandments, 142; *in Bible*
thalassophilia (sea-lust), 137
Thaxton, Charles, 141
theistic evolutionists, 129
theorems, 6–7
theos (god), 57, 98; *in Bible*
Thingvellir National Park, Iceland, 78
Thompson, D'Arcy, 23–24, 42–43, 62
Thompson, Francis, 88, 95
Thompson, William, 85
tidal fluctuations, 81
Tigress River, 134
tomato plants, 9–10, 15, 24–25, 48
Tombaugh, Clyde, 33
tongue stone, 110
transcendentalism, 32
treatment, 9, 15, 46, 48, 66
Truman, President Harry, 22
Turing, Alan, 37

uniformitarianism, 76
United States, 7, 31, 137
universe, 17, 29, 33, 54, 72–73, 75, 78, 87, 94, 96, 99, 104, 124, 127, 140
    age of, 55–56
    beginning of, 1, 3, 11, 17, 34–35, 53–54, 56–58, 60, 72, 96, 115
    center of, 74
    dynamics of, 59, 97
    expanding, 17–18, 54–55, 119
    formed by chance or design, 57, 96–97
    multiple, 58
    open or closed, 125

Subject Index

suitability for life, 96, 106
University of Hawaii, 64
University of Oregon, 19, 23
University of Washington, 43
Ussher, Archbishop, 83

Vigeland, Gustave, 142
viruses, 107–8
volcanoes, 76

Wagner, Moritz, 108
Wallace, Alfred, 108, 113
war, 3, 6, 29, 33, 41, 59, 78, 141
    between science and religion, 124, 130
    war within Christianity, 127
    Cold War, 78
    Korean War, 2
    unfit for war, 45
    U.S. Civil War, 134
    Vietnam War, 41–42, 45
        Gulf of Tonkin affair, 19–20; Tet Offensive, 42
    World War II, 1, 2, 4, 17, 19, 23, 28, 41, 43–44, 50, 78, 97, 131
"warm little pond," 103–4; see Darwin, Charles

water, 2, 41, 47, 50, 75–76, 80–82, 98, 101–2, 107, 112, 123, 129, 140, 142
Watts, Los Angeles, California State, 133
Wegener, Alfred, 17, 77
Weismann, August, 116
Weltanschauung, 27; see worldview
whales, 23, 43–44, 71
White, Andrew D., 124
White, Ellen G., 81
Willamette River, 40
Wilson, Robert, 54
Wilson, Woodrow, 136
Wittenberg, Germany, 68
Wittgenstein, Ludwig, 54
Woit, Peter, 58
worldview, 17, 27–34, 38–39, 60, 104, 125

Y-chromosome, 135
Yahweh (Lord), 97, 139; *in Bible*
Ying and Yang, 58
Yom (day), 75, 83–84, 86, 107; *in Bible*
Yucatan Peninsula, Mexico, 111

Zechariah, 55; *in Bible*
zero, 7, 55

Scripture Index

Genesis

Book	17, 36, 53, 57–58, 73, 75, 84, 101, 110, 129–30, 133–34, 140, 158
1:26	119
1:22	107
1:14–18	73
1:9	76
1:7	74
1:3–5	75
1:3	73
1:1–2	75
1:1	53
2	139
2:23	141
2:14	134
3:2–32	83
5	83
6	80
6:12	80
7:18–19	80
7:22	80
9:20	80
11:4	134
13:13	74
29	109
32:28	156
41:57	81

Exodus

3:13–15	139
15:5	76

Leviticus

17:11	141

1 Samuel

2:8	74

1 Kings

4:34	81
19:11	76

2 Kings

Book	83
23:5	73

Job

9:9	74
9:8	55
33:4	140
36:27–29	129

Scripture Index

Psalms

8:3	30
93:1	74, 129
104	129
104:3	58
113:3	74
114:3	76

Proverbs

8:22–3	57
25:2	156

Ecclesiastes

1:6–7	129

Isaiah

40:22	129
42:5	55

Jeremiah

9:1	142
33:22	129

Daniel

4:10–11	74, 129

Amos

5:26	74
9:6	75
9:5	76

Micah

1:4	76

Zechariah

12:1	55

Matthew

Book	83, 97
3:17	98

Mark

	97

Luke

	97
2:1	81

John

Book	97
1:32–34	98
1:1	53, 57
8:32	12
13:34	143
15:13	143
18:19	151

Acts

Book	70, 147
4:13	151
5	70
9	95
16	95
28:2–7	110

Romans

1:20	156
1:19	141
12:2	152

Scripture Index

1 Corinthans

13:2   153
16:19   70

Galatians

5:1   100

Ephesians

2:5–10   100
4:25   12

www.ingramcontent.com/pod-product-compliance
Lightning Source LLC
Chambersburg PA
CBHW062044220426
43662CB00010B/1650